Building and Civil Engineering Construction

Volume 2

Building and Civil Engineering Construction

Volume 2

BRIAN W. BOUGHTON
C.Eng., M.I.Struct.E.

Advisory Officer for
Building, Civil Engineering and Surveying,
Technician Education Council

GRANADA
London Toronto Sydney New York

Granada Publishing Limited – Technical Books Division
Frogmore, St Albans, Herts AL2 2NF
and
36 Golden Square, London W1R 4AH
515 Madison Avenue, New York, NY 10022, USA
117 York Street, Sydney, NSW 2000, Australia
60 International Boulevard, Rexdale, Ontario R9W 6J2, Canada
61 Beach Road, Auckland, New Zealand

ISBN 0 246 11967 5

First published in Great Britain 1983 by Granada Publishing

British Library Cataloguing in Publication Data
Boughton , Brian W.
    Building and civil engineering construction.
    Vol. 2
    1. Building
    I. Title
    690   TH145

Typeset by Cambrian Typesetters, Aldershot, Hampshire
Printed in Great Britain by Richard Clay
(The Chaucer Press) Ltd, Bungay, Suffolk

Granada ®
Granada Publishing ®

Contents

This book is intended to cover the content of the TEC units
  Building Construction III,
  Civil Engineering Construction III,
together with elements of
  Materials in Construction II.
In developing the earlier coverage of Volume 1 the first chapter will
also cover some elements of
  Organisation and Procedures II,
  Site Surveying and Levelling II, and
  Structural Detailing II.
As with Volume 1, the material will be considered in an integrated
form, with each chapter including some assignment work. Reading of
digests, technical journals and publications relating to specification
and construction of building and civil engineering work is encouraged,
both as a means of assisting in the assignments and as a continued
discipline for updating of technical information. Whereas Volume 1
was concerned mainly with developing the reasoning of 'how and
why', this book is aimed at promoting the logical reasoning more
fully into 'where and when', with consolidation of the 'how and
why' aspects.

1 Forms of Construction

In Volume 1 we looked at some of the more usual forms of construction from the contractor's point of view. In practice, a proper analysis of the choice of materials, techniques and construction forms must balance the interests of all the major contributors. The people involved will, by virtue of their background, quite often place a different emphasis on the importance of the various facets of the construction. Our aim in this chapter is to attempt to balance these aspects to produce solutions that best satisfy all of their interests.

Obviously, every contract is different in some way and it would be irrational to think that we can cover all eventualities. If, however, we can begin to reason logically as to why one answer might be better than others, the development and consolidation of such reasoning will hold us in good stead when we are placed in a situation that has some similarities to one of those considered here. Background reading of technical journals and magazines reporting on 'real' contracts can also help to consolidate the analytical approach, as can an active interest in the construction going on around us all the time. So that the various interests can be considered, we shall look at alternatives for all aspects of the construction in some case studies of low-rise domestic, commercial and industrial forms of construction.

CASE STUDY 1.1

A local-authority housing estate is to comprise a mixture of terraced two- and three-storey housing, flats and maisonettes. Garaging is to be provided in blocks so that access to the houses is by pedestrian walkways only; service roads are, however, required for delivery vehicles and furniture-removal lorries.

The first consideration of the designer (architect/building surveyor) is space utilisation. How can the space provided be best used to accommodate the families, produce an attractive estate layout and offer areas for recreation and leisure? As far as space related to people is concerned there are two limiting factors:

(a)   the number of people per hectare (or per acre), this being the 'density of population'. This is fixed by the planning authorities and is subject to small variations between suburban and city areas nationally.

(b)   the floor area requirements per person per dwelling, this being the 'space in the home'. The space requirements will vary, depending both on the type of house, flat or maisonette and on the number of people (i.e. bed-spaces) for whom it is designed. These parameters were, at one time, laid down by central government in guidelines to local authorities (many of whom still work to them) and include storage space requirements. Examples are shown in fig. 1.1.

Once these factors are satisfied, the designer has freedom to lay the estate out in such a way as to make the best use of its topographical nature (i.e. slope of ground, trees and bushes, streams, etc.). He will, of course, still need to satisfy planning requirements that the estate is visually acceptable. The layout design will include the siting of the parking areas to allow for short pedestrian travel to the dwellings, the footways and the access roads (e.g. for postal and milk deliveries). (See fig. 1.2.)

Coupled to the overall visual layout is the appearance of the individual dwellings. Although the style and colour will have some bearing, the major effect will be created by texture and the choice of materials. It is at this stage that the views of other members of the team need to be considered.

Materials

The designer will be concerned primarily with appearance and durability. These two factors combine in determining the weathering characteristics as far as external surfaces are concerned. However, they are of individual importance in relation to internal surfaces.

The structural engineer will be concerned primarily with stability and structural performance. It may be necessary to use supplementary materials to achieve these targets but economically it is more sensible to utilise those serving the main functions.

The builder will be more interested in the handling, storage and fabrication of the materials. He will need to establish that his workforce has the necessary experience of the use of the materials to

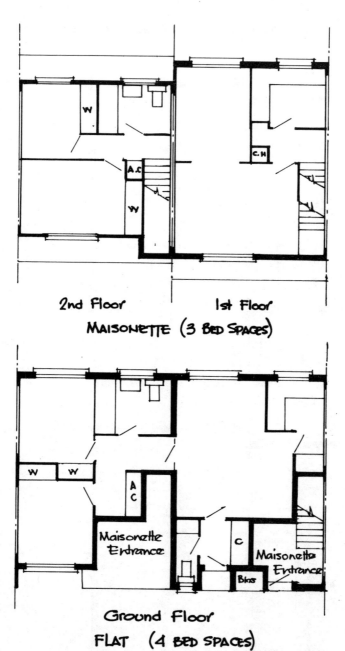

2nd Floor   1st Floor

MAISONETTE (3 BED SPACES)

Ground Floor

FLAT (4 BED SPACES)

Requirements for Min floor area (incl st)	3 B.S.	4 B.S.	5 B.S.	6 B.S.
	62m²	80m²	90m²	98m²

Fig 1.1   LAYOUTS TO PROVIDE FLOOR
AREAS IN EXCESS OF MIN. REQU'TS

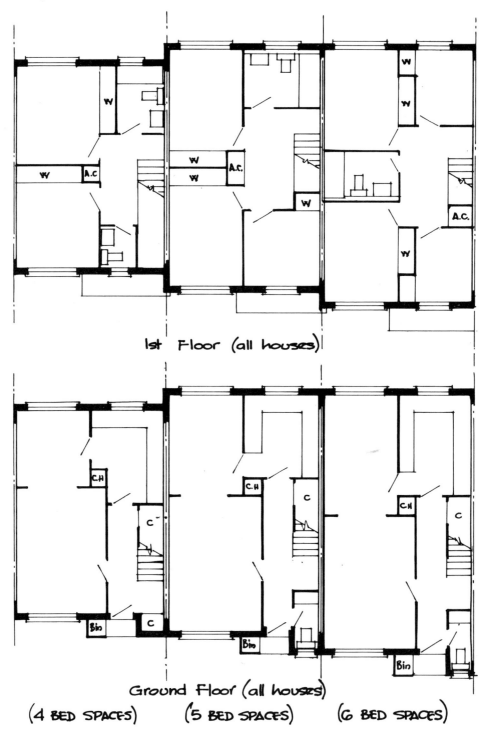

1st Floor (all houses)

Ground Floor (all houses)

(4 BED SPACES)   (5 BED SPACES)   (6 BED SPACES)

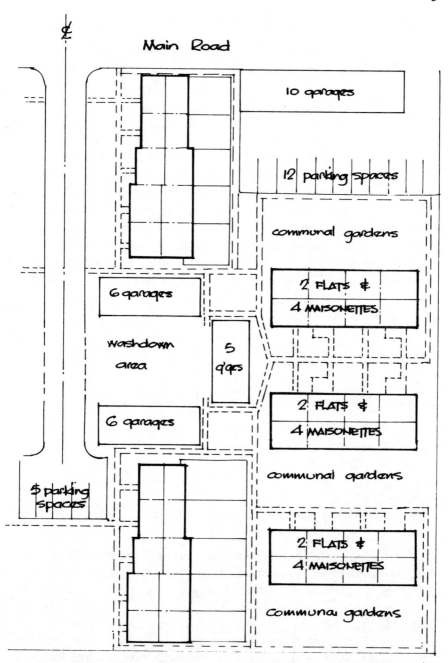

Main Road

10 garages

12 parking spaces

communal gardens

2 FLATS &
4 MAISONETTES

6 garages

washdown
area

5
g'ges

2 FLATS &
4 MAISONETTES

6 garages

communal gardens

5 parking
spaces

2 FLATS &
4 MAISONETTES

communal gardens

FIG 1.2 SHOWING HALF OF A SYMMETRICAL SITE
LAYOUT OF MIXED HOUSES, FLATS & MAISONETTES

produce adequate standards without undue extra expenditure on his part.

'Building Control' will also be interested in stability and standard of construction, but these members of the team are concerned additionally with fire risk and flame spread. This aspect will also be the responsibility of the designer in selecting appropriate materials. Suitability and performance will therefore need to meet the requirements of the building control or district surveyor's department.

The quantity surveyor will be concerned almost exclusively with the economic performance of the materials, i.e. their basic cost, the labour necessary to produce or install them, and their maintenance and replacement costs. He will, of course, need a background knowledge of the other aspects already outlined in assessing their 'value for money' when making recommendations to the client.

Technicians working in all the categories listed here should possess an informed background in the other categories in order to converse adequately with each other. This being the case, let us look at the sensible choices available.

(1) *Roofing.* This will depend to a large extent on the final choice of roof form since flat roofing can utilise materials that differ from those appropriate to pitched roofing. If a flat roof is used, the finishing materials that might be employed are:

(a) built-up felt;
(b) asphalt;
(c) zinc sheeting;
(d) asbestos cement tiles, which might be used in conjunction with (a) or (b);
(e) lightweight aggregate bound in bitumen, which might be used in conjunction with (a) or (b).

Built-up felt is liable to deterioration through temperature changes that might cause brittle fracture. Water may then get trapped, either under or between layers where it may bubble in hot conditions or freeze in cold conditions. In both cases this leads to further breakdown of the material, which can allow water penetration to the supporting structure. Unless protected, built-up felt is also likely to be damaged by pedestrian traffic. Although this is normally only likely to occur during maintenance, it might happen because of children climbing on to the roof to retrieve footballs, etc. or as part of a game or dare.

Asphalt also suffers from extremes of temperature, becoming soft in warm weather and brittle in cold conditions, and is likely to bubble or fracture. In addition it is likely to 'drift' in warm winds so that some areas of the roof become thinner than others. This can accelerate its deterioration.

Zinc sheeting is not so likely to deteriorate as a result of temperature changes but it may be affected by chemicals contained in the atmosphere. This applies particularly in industrial locations, which might be adjacent to a housing estate of this nature. [*Note*: Lead sheeting has been ignored deliberately because of the high risk of its being 'stripped' for sale as scrap metal, its price being sufficiently attractive to tempt such action. It is possible that some potential young 'villains' might mistake the zinc sheeting for lead so this becomes an additional consideration.] Zinc sheeting is rather more expensive than built-up felt or asphalt and requires more expertise in its application. This might therefore make it economically unsuitable for consideration.

The choice would appear to rest between (a) and (b), but since both suffer fairly rapid breakdown when exposed it is sensible to protect them in some way as a means of prolonging their useful life.

Asbestos cement tiles protect the surface from pedestrian traffic and provide a reflective surface to reduce the effect of hot sunlight. They also give a small amount of thermal insulation. However, they are unlikely to reduce the effect of cold weather causing brittle fracture, either to built-up felt or asphalt.

Bituminised lightweight aggregate provides a greater depth of protection and gives good thermal insulation to reduce temperature effects on the materials. It also deters pedestrian traffic since it provides a less stable footing and sticks to shoes on contact.

If, therefore, a logical choice is to be made, it would most probably be asphalt topped with at least 60 mm of bituminised lightweight aggregate.

If a pitched roof is used, the finishing materials that might be employed are:

(a) built-up felt, reinforced with expanded copper sheet;
(b) ribbed aluminium;
(c) plain tiles, clay or concrete;
(d) semi- or fully interlocking concrete tiles.

Built-up felt will require a sub-grade of ply or wood-wool sheeting and will need careful supervision to achieve an acceptable finish. The joints will need to align vertically and there should be no ripple or bubbling of the layers. If a good finish is achieved it can be quite pleasing visually, since the expanded copper will gradually colour the roof green. Felt does not, however, last as long as tiles and since its initial cost is unlikely to be lower than that of tiling it is not the obvious choice.

Ribbed aluminium sheeting does not require the degree of support demanded by the other materials but has some drawbacks. It is extremely noisy if birds walk about on the roof; it is also rather bright and could cause irritating reflections in sunlight. From a cost consideration it is probably no more expensive than tiling but might deteriorate more quickly in a salty atmosphere (this is not likely in an inland location).

Plain roof tiles require a steep roof pitch, which would mean increased roof space. This might be useful as storage but requires more space heating. The durability of tiles is proven as being equal or superior to most other pitched roof finishes, the exception being slate, which is rather expensive initially.

Interlocking tiles have very good weathering characteristics, require a lower roof pitch and, because they only require single lapping, present less dead load on the roof structure. Therefore, if pitched roofing is chosen, the most suitable finish would seem to be inter-locking concrete tiles.

(2) *Roof structure.* As far as the choice of structural material is concerned there would appear to be no great argument. Although steelwork, reinforced concrete and prestressed concrete can achieve the strength requirements, timber is much easier to install. Timber sections are also lighter to handle, more workable if modifications are required and more readily understood by the work-force. The choice is in the structural form to be adopted, which will be governed by the selection of flat or pitched roof construction.

For *flat roofing* the roof joists can be either solid or expanded, depending on the span between supports. However, since the span is dictated by room sizes, it is logical to use the internal walls as support. This means that solid joists can be used economically. [*Note*: If plasterboard partitions were to be used for the internal walls, it would be worth considering 'box beams'. This is not, however, a sensible consideration for local-authority housing — see later text.]

In *pitched roofing* the choice will depend to some extent on the use to which the roof might be put (storage or partly habitable) and also on the supporting structure.

(a)   trussed purlins and common rafters;
(b)   trusses, purlins and common rafters;
(c)   trussed rafters, conventional;
(d)   trussed rafters, open;
(e)   mono-pitch trusses, purlins and rafters;
(f)   trussed mono-pitch rafters.

Trussed purlins can be usefully employed in 'cross-wall' construction (as described in Volume 1 — case study 5.2) and provide open central space. The roof space behind the purlins is, however, unusable.

Trusses, purlins and common rafters use rather more timber than modern roof forms and are also labour intensive.

Conventional trussed rafters are quick to install and, being factory produced, are constructed to a good degree of accuracy. However, they do tend to reduce the amount of usable roof space.

Open trussed rafters combine the benefits of normal trussed rafter construction while adding the value of a large usable area of the roof space. This allows for habitable rooms to be accommodated in the roof space, thus improving the efficiency of the building.

Mono-pitched roofing may be used for the general roof area of buildings. However, it is more suitable for small areas of roofing at lower levels as an alternative to flat roofs. Similar economic arguments may be made in relation to trussed rafters against trusses, purlins and rafters as have been stated for full trussed roofs.

Since pitched roofs in general cause fewer problems than flat roofs, the choice of construction should be for open trussed rafters generally, with mono-pitched trussed rafters at the lower levels. The exception is where flat roofing may be dictated by planning considerations.

Regardless of the type of construction adopted, all components should meet British Standards requirements, and must form a structurally stable roof. The roof must be dimensionally co-ordinated to the tiles being used (to avoid the need for cutting). The timber must have the correct moisture content, and should be pressure impregnated with preservative. The minimum thermal insulation requirements should be exceeded. Depending on where the thermal insulation is located, a vapour barrier and/or roof ventilation must be provided.

(3) *External walling*. This is sometimes known as the building envelope (the roof also being included) and is in many respects one of the most critical parts of the structure. Its visual treatment must conform to planning requirements, it must subscribe to the environ-

ment and provide an attractive setting. The materials used must be durable, they should require little or no maintenance and must provide an effective barrier to adverse weather conditions. The external walling must provide good thermal insulation and, depending on its construction, this may affect the window:wall ratio. It should also afford good sound insulation.

The forms of external wall construction for housing fall into three main categories:

(a)   solid − either brickwork, stone or concrete;
(b)   cavity − brick and block or alternative outer-leaf materials of stone or concrete;
(c)   stud timber − either timber frame or curtain-wall infill.

Solid walling is good as a sound insulator but does not perform well in the exclusion of driving rain unless rendered. Since the rendering may move and crack at some stage, there will always be the long-term danger of damp penetration. Except for aerated concrete blockwork, solid walling is usually poor in its performance as a thermal insulator; dense concrete, stone and brickwork create problems of condensation within the wall thickness or on its inner surface. One way to overcome the problems of weather penetration on the solid aerated block wall is to provide an external surface that is raised forwards of the wall on battens, e.g. tile-hanging or weather-boarding (as described in Volume 1, chapter 4) with a barrier of building paper or felt at the back of the battens. This surface is not very stable against impact forces and, if used in local-authority housing, should be restricted to upper walls.

Cavity walling performs reasonably well as a sound insulator and any rain penetration is stopped at the inner face of the outer leaf of brick, block or stone by the cavity. If the cavity is 'bridged', however, the damp may track across to the inner leaf. This should be avoided by ensuring that (i) mortar is not dropped into the cavity or on to the wall ties during construction, and (ii) any cavity insulation does not form such a 'bridge'.

Although cavity insulation can be 'blown in' after the wall has been completed, it is more sensible to produce a 75 mm cavity with 50 mm of rigid insulation to the inner leaf, thus retaining a 25 mm gap to allow rain-water to drain or evaporate on the outer leaf. It is possible to meet the limiting thermal insulation values by using very lightweight aerated blocks without insulating the cavity. However, it is sensible to improve the rating above the minimum to reduce heating costs, particularly since these will continue to rise with inflation. The finish to the inner wall may be dry-lined or plastered. However, although dry lining or lightweight plaster may further improve the thermal insulation, neither of these is as durable as dense plaster. Durability is an important consideration in public-sector housing. An added benefit of dense plaster is its contribution to sound insulation.

Timber studwork can accommodate 100 to 150 mm of glass-fibre insulation, depending on stud sizes, which contributes to a very good thermal insulation property. It is not, in itself, a very good sound insulation medium and will depend on the external surface treatment to improve this factor. An outer leaf (or veneer) of brick, stone or concrete block is the ideal method for producing the required reduction in airborne sound, since such a surface will also perform other functions. The internal finish of dry lining, while perfectly acceptable in private-sector housing, does not give the level of resistance to impact force of dense plaster. Some improvement in this regard can be achieved by using the reverse side of the wallboard and a skim coat of dense plaster. For halls and landings, however, plywood panelling should be preferred.

The choice of external walling is, by necessity, a compromise. Whereas private-sector housing, particularly that built to specific customer requirements, would choose timber-framed construction (true to shape, no drying-out problems, excellent thermal insulation), the choice for local-authority housing might well be cavity wall construction (more durable against heavy treatment internally). The final decision might, however, be influenced by the time scale and labour component of the construction. This is because the construction period, the drying out prior to decoration and the making good after occupation (shrinkage cracks, etc.) take much longer for cavity walling than for timber-framed work.

As far as external finishes are concerned, these will be varied, being brick at ground floor and a selection of brick, vertical tile hanging or u.P.V.C. weather-boarding at first floor and at gables to second floor. All of these require minimal maintenance and to avoid the monotony of repetition the shades of brickwork, tiling and weather-boarding can be varied within basic colour ranges. (See fig. 1.3.)

(4) *Windows and doors.* The range of windows available, both in materials and in styles, is vast. How, then, can we set about making a logical selection? First, we can discount the consideration of

MIXED TERRACE OF 2,3 & 4 BEDROOM HOUSES

MIXED TERRACE OF FLATS & MAISONETTES

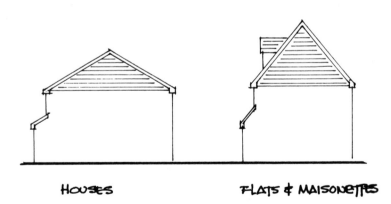

HOUSES   FLATS & MAISONETTES

FIG 1.3 SHOWING TREATMENT TO ELEVATIONS
     TO PRODUCE VARIATION OF FINISH

purpose-made frames, regardless of material, since the costs by comparison with their counterparts within standard size ranges is doubled or trebled. Secondly, we should consider only those that can accept sealed-unit double glazing, not only because of the reduction in heat loss but also because of the reduced risk of condensation. This will eliminate a large proportion of the range, e.g. modified B.S. softwood windows, standard metal sash windows, etc. Thirdly, we should consider decorative maintenance, looking to those that require a minimum of attention. This highlights the value of (a) aluminium frames, which require no painting, and (b) hardwood frames, which require periodic preservative stain. There is a third alternative of u.P.V.C. reinforced with steel plate. However, this material does not perform well in fire conditions, having a low melting point, and can be discounted. A fourth point for consideration is functional maintenance. This is particularly important in local-authority housing, and will reduce the choice generally to top- or side-hung windows with peg stays. The exception to this would be for windows in the roof space, if aligned with the roof pitch, which would by preference be pivoted. In general, however, we can eliminate horizontal and vertical sliding sashes, horizontal and vertical pivot windows, friction hinges and composite action pivot/ side-hung frames, since all of these could require potentially more maintenance than the simple hinged and stayed frames.

Although double glazed windows are more expensive than single glazed, the extra expense can be recouped quite quickly in running/ maintenance costs, which, in addition to heating, also include internal decoration. The condensation that occurs on single glazed windows can affect the reveals and window boards and (in the case of timber windows) the frames, causing dampness in the plaster, the risk of rot in timberwork and general discoloration of internal decoration. If aluminium frames are selected, they should incorporate a 'thermal break' to avoid the frame acting as a conductor and forming a 'cold bridge' with consequent condensation on the inner face.

The final choice will probably be dictated by costs: little or no maintenance of the aluminium frames but only of the hardwood sub-frames, or no risk of internal condensation offset by periodic treatment to hardwood frames. On balance the hardwood might be chosen since it will blend better with 'natural' finishes of brick and tile hanging. It is also likely that small dormers may be preferred for windows in the roof space but both might be used to gauge the long-term cost-in-use figures for both.

As with windows, the range of doors available is considerable. However, by far the most common type of entrance door is of panelled softwood in a softwood frame. To retain the character of the elevations this type of door would probably be chosen even though it would require painting at 3- to 4-year intervals. Its main virtue is that it can readily accept all sizes of lock, and can accommodate additional security locks, bolts, chains, etc., which can be fitted by the tenants.

(5) *Internal walling.* This may be loadbearing or non-loadbearing. The former is available in either brickwork, blockwork or timber studwork, and the latter in all of these and additionally in lightweight plasterboard partition work. Because sound insulation, particularly between dwellings, is important, and to allow for dense plaster finishes, the most likely choice is blockwork. If, however, timber-framed construction is adopted, the detail at party walls has to incorporate a sound deadening quilt and the dry lining is applied in two thicknesses with staggered joints. As well as reducing the risk of airborne sound passages the stagger improves the resistance to attack and spread of fire.

(6) *Suspended floors.* For single dwellings, i.e. two- and three-storey houses and upper floors of maisonettes, timber joisted floors, decked in tongued and grooved chipboard or plywood and soffitted with plasterboard, are the obvious choice. They are light and easy to assemble; the materials are easily workable; and provision for internal services causes no problems. For multiple dwellings, however, the separating or 'compartmenting' floor must resist fire spread. As this floor should also resist the passage of sound it is advisable to use a dense material. Concrete is the ideal choice, being incombustible and possessing fairly good sound insulating qualities. The construction may be precast or in situ concrete reinforced with steel bars. (Precast concrete construction was described in Volume 1, case study 5.2.) To simplify the construction of in situ r.c. floors, the supporting structure can be of wood-wool slabs, which act as 'permanent formwork'. This means that the slabs are not 'struck' (removed), as would be the case with plywood, but are retained to provide the floor soffit. This technique offers a few advantages over the precast plank floor system, namely:

(a)   the wood-wool improves thermal insulation;
(b)   the soffit accepts plaster more easily;
(c)   service conduits can be placed on the formwork prior to fixing the reinforcement or pouring the concrete;

(d)   there is no need for cranage;

(e)   the in situ slabs provide better structural stability.

As with the precast planks, the upper finish may be screeded but is better served by using a 'floating floor'. A layer of rigid polystyrene sheeting topped by tongued and grooved chipboard or plywood serves to reduce impact sound from above and provides an ideal surface on which to lay tiles or carpets. Services for the upper dwelling can also be accommodated in the sandwich layer of expanded polystyrene. It is, of course, important to ensure that support of the shutters is retained for at least 14 days after the concrete is poured so that the finished floor can attain its design strength.

(7)   *Ground slab and foundations.* The oversite concrete, acting as ground slab, is an obvious choice for ground floor construction because it is more economic than suspended timber flooring. A 'floating floor' finish similar to that for the maisonettes is to be provided, with utility areas given a screed and quarry tiles. The foundations are, of course, dependent on the ground conditions, which in this particular case show a large proportion of soft clay to a depth of about 1.5 m. This will produce considerable seasonal movement and the type of foundation least likely to cause problems is a combination of short bored piles topped by ground beams. A compressible packing will need to be located at the underside of the ground beams to accept swelling of the subsoil in winter conditions without producing upward loading on the beams. [*Note*: Drainage runs should be laid to greater falls than normal to avoid reverse falls occurring in wet ground conditions.]

Summarising the choice of construction, based on logical consideration of all factors, the buildings are to be of 'cellular' construction in brick and blockwork using cavity walls to all perimeter walls. Flooring is to be solid at ground floor level, in situ reinforced concrete to dividing floors between flats and maisonettes, and timber joists elsewhere. Windows are to be of hardwood construction, double glazed, and upper wall finishes will incorporate tile hanging and weather-boarding elements. Roofs are to be of open trussed rafter form with interlocking tiles to utilise part of the roof space. To keep maintenance costs to a minimum, it is decided to use 'off-peak' electric storage heating in combination with a very high level of thermal insulation. This eliminates the need for flues, pipe runs, boiler servicing, room thermostats, etc. associated with more complex systems of heating. Hot water is provided by 'white meter' immersion heaters coupled to large, well-insulated storage cylinders. Details of the elements of the construction are shown in figs 1.4 to 1.7.

CASE STUDY 1.2

A speculative office building comprises two three-storey office blocks serviced by a central 'core' housing stairs, lifts and toilet facilities.

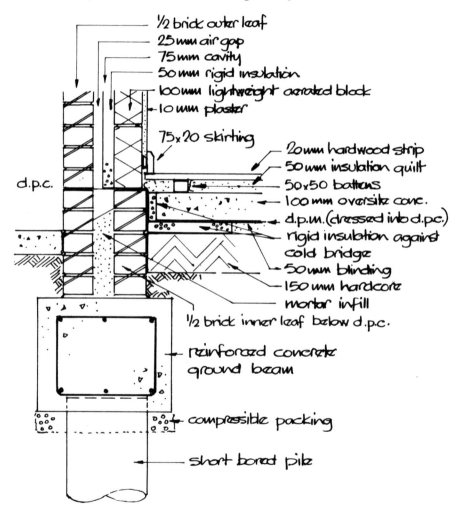

½ brick outer leaf
25 mm air gap
75 mm cavity
50 mm rigid insulation
100 mm lightweight aerated block
10 mm plaster

75 x 20 skirting

d.p.c.

20 mm hardwood strip
50 mm insulation quilt
50 x 50 battens
100 mm oversite conc.
d.p.m. (dressed into d.p.c.)
rigid insulation against cold bridge
50 mm blinding
150 mm hardcore
mortar infill
½ brick inner leaf below d.p.c.

reinforced concrete ground beam

compressible packing

short bored pile

Fig 1.4 DETAIL SHOWING 'SHORT BORED PILE' FOUNDATION, INSULATION AT GROUND SLAB TO PREVENT COLD BRIDGE, AND WALL CONSTRUCTION

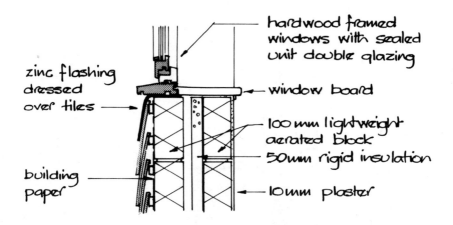

zinc flashing dressed over tiles

hardwood framed windows with sealed unit double glazing

window board

100 mm lightweight aerated block

50mm rigid insulation

10mm plaster

building paper

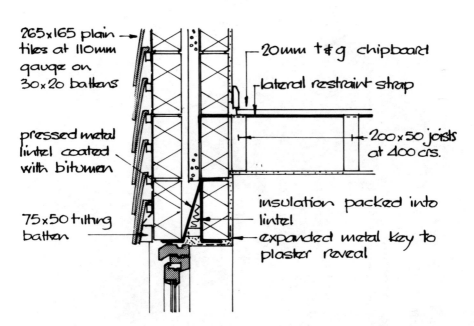

265x165 plain tiles at 110mm gauge on 30x20 battens

20mm t&g chipboard

lateral restraint strap

pressed metal lintel coated with bitumen

200x50 joists at 400 crs.

75x50 tilting batten

insulation packed into lintel

expanded metal key to plaster reveal

Fig 1.5 DETAIL THROUGH EXTERNAL WALL OF HOUSES SHOWING CHANGE FROM BRICK OUTER LEAF, TO TILE HANGING ON BLOCKWORK, AND PRIMARY DOUBLE GLAZED WINDOWS

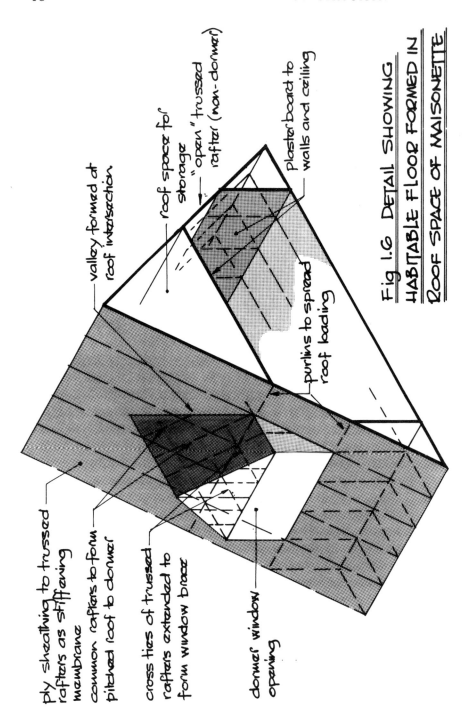

valley formed at roof intersection

roof space for storage

"open" trussed rafter (non-dormer)

plasterboard to walls and ceiling

purlins to spread roof loading

ply sheathing to trussed rafters as stiffening membrane

common rafters to form pitched roof to dormer

cross ties of trussed rafters extended to form window brace

dormer window opening

Fig 1.6 DETAIL SHOWING HABITABLE FLOOR FORMED IN ROOF SPACE OF MAISONETTE

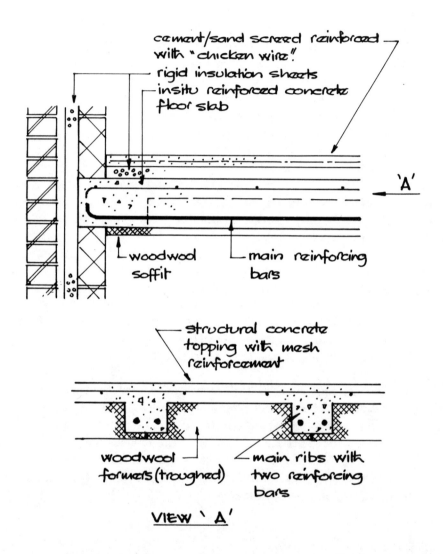

cement/sand screed reinforced
with "chicken wire"
rigid insulation sheets
insitu reinforced concrete
floor slab

`A'

woodwool
soffit

main reinforcing
bars

structural concrete
topping with mesh
reinforcement

woodwool
formers (troughed)

main ribs with
two reinforcing
bars

VIEW `A'

Fig 1.7 DETAILS OF RIBBED INSITU
R.C. FLOOR BETWEEN FLATS AND
MAISONETTES PRIOR TO PLASTERING

Parking is provided below the offices at an open ground-floor area either side of the 'core'. The offices are to be of open-plan format but must be capable of supporting partitioning as required.

The first consideration of the designer is optimum space utilisation — how to make the best use of a known width and depth of site. How can the service core be kept to a minimum width that will afford access to stairs, lifts and utilities? Circulation space becomes important here since it must be sufficient to cope with the number of people envisaged for the offices.

As far as space requirements (office areas) are concerned, these are outlined in British Standards, for both individual and integrated offices. As the building is 'speculative' (not built to a specified requirement), it would be sensible to assume an overall space per head. A useful figure to work on would be 12 m^2 per person, excluding the service core. This allows for some flexibility since a typing pool might drop to 5 m^2 per person whereas a director might require 30 to 35 m^2 of floor space. For the layout shown in fig. 1.8 this is approximately twenty persons per floor per wing. It would be sensible therefore, to provide two eight-person lifts (or even ten-person, if possible), with approximately 24 m^2 of circulation/waiting space at the upper floors (sufficient for both office wings).

The logic in arranging the lifts, stairs and toilets as shown is that:

(a)   the lift motor room will be to one side of the service core;
(b)   ducts can be provided behind each lift shaft to house fire risers (stair access panel) and sanitation pipes (from toilets);
(c)   fire escape stairs are to the rear of the service core;
(d)   since toilets are not required at ground floor level (except for a single toilet for reception), this can provide extra space for the main entrance/reception area.

Materials

The designer must consider fire safety, flexibility of use, visual acceptability and ease of maintenance. These factors determine the structure and cladding material selections that will also prove economic to build.

The structural engineer will be concerned with stability and performance but will also advise on the most suitable structural form to produce an economic solution.

The services engineer will be interested only in the mechanical

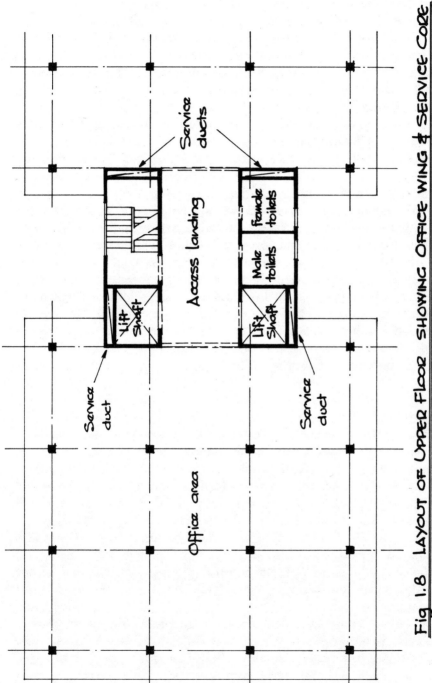

Fig 1.8 Layout of Upper Floor showing Office Wing & Service Core

services for the central core, and the facility for power, lighting, communications and heating/ventilation to the office wings.

The builder will need to be familiar with the constructional techniques appropriate to the requirements of the designer, structural engineer and services engineer and will need the ability to exercise full quality control in the selection, storage and erection of the materials and components.

'Building Control' will also be concerned with stability, quality and an understanding of the construction techniques. They are involved additionally in fire risk, flame spread and means of escape. In addition to these aspects the sanitation performance must meet the requirements of the regulations or by-laws.

The quantity surveyor will be concerned very closely with the economic performance of the materials and must be familiar with their associated constructional techniques. This is so that he can advise the client and control costs and stage payments.

Technicians working in all the above categories must have some background knowledge of the associated categories in order to avoid delays, misunderstandings and disputes, which can disrupt the smooth running of the contract. This being the case let us analyse the reasons for the choice of materials, structural form and claddings, finishes, etc. adopted for this particular building.

Structural form

This is directly related to the choice of materials and cladding. Steel-framed construction can be discounted because of fire protection requirements, the risk of structure-borne sound and the requirement for downstand beams, which would affect office flexibility. It would also be unsuitable for the service core. Precast concrete would not provide easy accommodation of the office-block services and would also suffer from the need for downstand beams. Reinforced concrete 'framed' construction also employs downstand beams. The logical choice is therefore *flat-slab construction* utilising columns in the office wings and walls in the service core for its vertical support component.

Flat-slab construction has several advantages in the design of office blocks. It provides a flat soffit, which allows the use of demountable partitions, giving maximum flexibility of layout for speculative rental. Additionally, its solid 'plate' floor is very strong and requires a thickness, in the case studied here, of only 125 mm, which reduces storey height. This has an effect on stair flights, cladding and the overall height of the building.

Since the material used is in situ reinforced concrete, it is also possible to reduce construction costs further, using well-proved current techniques. Table-form shuttering can be used for the soffit of the slabs; the conduit for power and lighting runs can be cast into the floor slabs with polystyrene 'blockings' at outlet positions; the floors can be 'power floated' to eliminate the need for screeding. It should also be possible to shutter the walls to the service core with sufficient accuracy that an applied decorative stipple is all the finishing required (the lighting conduit can also be located prior to casting the concrete). Unnecessary 'dead' loading is therefore kept to a minimum, producing a more economically functional structure. (See fig. 1.9.)

Cladding

In compliance with the concept of keeping structural dead loads to a minimum, the cladding dead loads are similarly restrained. Curtain walling has been chosen, using aluminium transoms and mullions sited inside the building, so that a smooth external surface is produced. Tinted glass is used to cill level with clear glass for the fenestration areas (windows). Timber studwork is provided internally to cill height to form a duct for the heating units. These run along the perimeter of both wings and are translated vertically in the service ducts. (See figs. 1.10 to 1.12.) Brickwork is used only for the walled sections of the central core. The wall ties are dovetailed into slots cast into the concrete, with an insulating sandwich of polystyrene sheet fixed to the concrete walls. This finish is carried through to the plant-room area over the service core to create a visual link. The cladding, both for the office wings and for the service core, continues above finished roof level to mask the parapet. (See fig. 1.13.)

Roof finish

The roof structure is similar to that of the floors but, instead of power floating to a finish, a lightweight screed, laid to falls, is applied. This is asphalted and then dressed with 100 mm of bituminised lightweight aggregate. This aggregate acts both as insulation against heat loss and as protection for the asphalt, which would otherwise be unsuitable for use. The effect of locating the insulation above the structural slab is to produce a 'warm' roof structure with minimal risk of condensation.

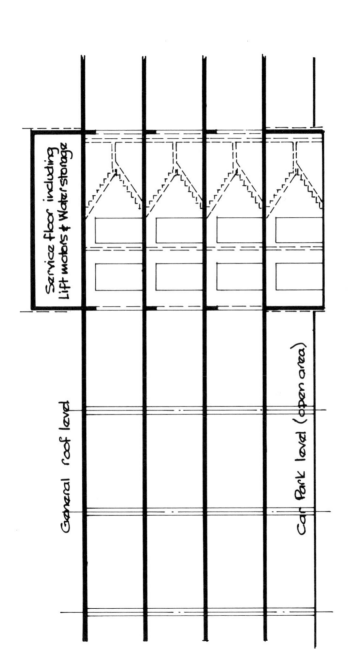

Fig 1.9 SECTION SHOWING OFFICE WING & SERVICE CORE

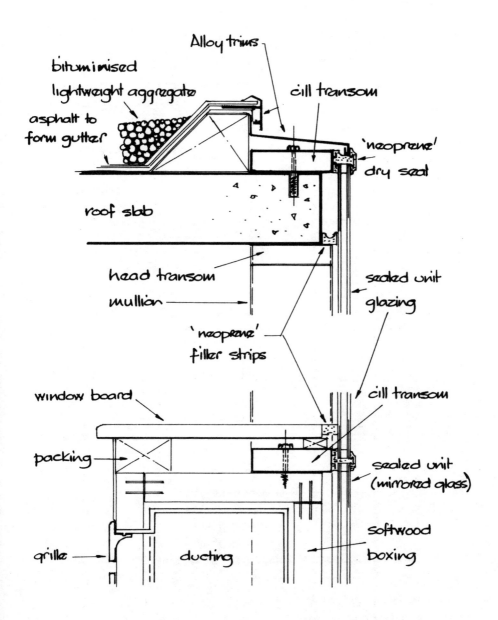

Alloy trims

bituminised
lightweight aggregate

cill transom

asphalt to
form gutter

'neoprene'
dry seal

roof slab

head transom

mullion

sealed unit
glazing

'neoprene'
filler strips

window board

cill transom

packing

sealed unit
(mirrored glass)

grille

ducting

softwood
boxing

Fig 1.10 Curtain Walling Details at Roof
AND CILL OVER A.C. DUCT (FIXED LIGHTS)

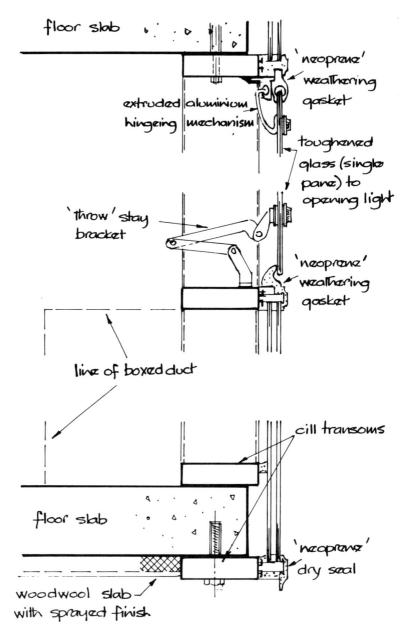

floor slab

'neoprene' weathering gasket

extruded aluminium hingeing mechanism

toughened glass (single pane) to opening light

'throw' stay bracket

'neoprene' weathering gasket

line of boxed duct

cill transoms

floor slab

'neoprene' dry seal

woodwool slab with sprayed finish

Fig 1.11 CURTAIN WALLING DETAIL AT FIRST FLOOR SHOWING SOFFITE & OPENING LIGHT

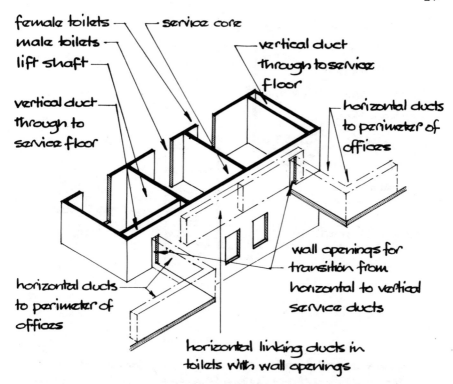

female toilets — ⌐ service core

male toilets —

lift shaft —

— vertical duct through to service floor

vertical duct through to service floor

horizontal ducts to perimeter of offices

horizontal ducts to perimeter of offices

wall openings for transition from horizontal to vertical service ducts

horizontal linking ducts in toilets with wall openings

Fig 1.12 SECTIONAL ISOMETRIC AT FRONT OF SERVICE CORE TO SHOW PROVISION OF HORIZONTAL AND VERTICAL DUCTING

Floor soffit to office wings

This forms the covered area to provide car-parking and will therefore need to house lighting. It would also be sensible to apply a finish that can provide insulation to the office floor above. This can be achieved quite easily by using wood-wool slabs as permanent form-work and spraying the underside with an expanded vermiculite plaster to a stippled finish. [*Note*: The cladding should provide a masked edge and restraining curb to the wood-wool and plaster. It would also provide protection from driving rain.]

Ground slab and foundations

The ground slab within the service core is of conventional 'solid ground-floor' form with the d.p.m. sandwiched between the blinding

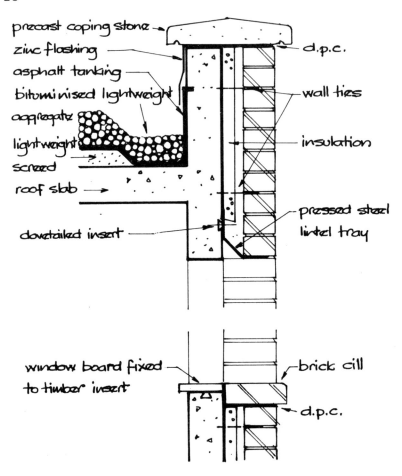

precast coping stone
zinc flashing
asphalt tanking
bituminised lightweight
aggregate
lightweight
screed
roof slab →
dovetailed insert

d.p.c.
wall ties
insulation
pressed steel
lintel tray

window board fixed
to timber insert

brick cill
d.p.c.

Fig 1.13 PARAPET DETAIL TO SERVICE
CORE WALL SHOWING CLADDING, ROOF
FINISH & TREATMENT AT WINDOW OPENING

and the oversite concrete. The junction with the walls is sealed with
a polysulphide joint to allow some degree of movement. A floating
floor finish is applied using polystyrene slab as insulation, topped
by a reinforced lightweight screed, and finished with quarry tiles.

The slab in the car-parking areas is provided with falls to allow for
surface-water drainage from the vehicles (melting snow), the drainage
system being located below. Access to the main drainage to the

offices is also located here, to the flank of the service core. Thermal movement joints are provided, to allow for variations in external temperature, on a 'chequer grid' system; concrete kerbs are located to produce defined entry/exit points. The surface is combed to assist in directing surface-water and to provide a skid-resistant surface.

An advantage of having kept dead loads to a minimum can be seen in the selection of the foundations. The heaviest load likely to occur on any column is not likely to exceed 600 kN and for the majority it is likely to be less than 450 kN. There is no need, therefore, to consider using piled foundations, since, for a ground-bearing capacity of 200 kN/m², the base 'plan' dimensions will not exceed 1.75 m². Also, because the 'formation' level of the footings will need to be around 1.2 m below ground level, there is no value in using reinforced spread footings, the thickness for mass footings being 750 mm maximum. Similarly, the wall foundations can be of mass concrete in wide strip form to spread the loads from the service core. (See fig. 1.14.)

CASE STUDY 1.3

A tractor components warehouse of plan dimensions 24 x 25 m requires a working height of 3.2 m for racked storage of variable-sized components. Gangway widths are to be 4 m to allow for the passage of fork-lift trucks and all natural lighting and ventilation is to be located within the roof space. Pallets are to be of 1.8 x 1.5 m plan size. The depths will vary in order to accommodate the different sizes of components but will fit into a standard racking system.

The designer has to consider a suitable floor layout to make the best use of the space provided and, in association with the structural engineer, to produce an economic framework that meets the specified requirements of the client. The racking dimensions to take the pallets can be approximated to 2.0 x 1.66 m to allow for tolerances. Since central racking can be in two banks, this will fit quite neatly into the 24 m plan (see fig. 1.15) in providing 4 m gangways. The racks can also be sited in banks of three to produce 5 m bays, which would fit the 25 m dimension.

As far as the constructional framework is concerned, it would be perfectly feasible to produce a perimeter of brickwork topped with a space-deck roof structure. However, the question needs to be asked: 'Is there any value in providing a completely clear internal floor space?' The short answer is 'No!' A large percentage of the floor space is taken up by static racking so that support columns can be

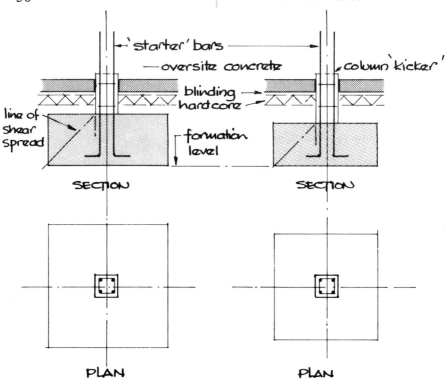

'starter' bars

oversite concrete

column 'kicker'

blinding

hardcore

line of shear spread

formation level

SECTION

SECTION

PLAN

PLAN

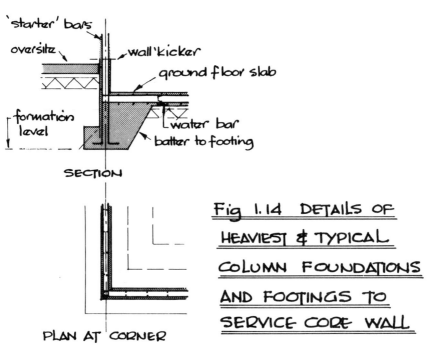

'starter' bars

oversite

wall 'kicker'

ground floor slab

formation level

water bar

batter to footing

SECTION

PLAN AT CORNER

Fig 1.14 DETAILS OF HEAVIEST & TYPICAL COLUMN FOUNDATIONS AND FOOTINGS TO SERVICE CORE WALL

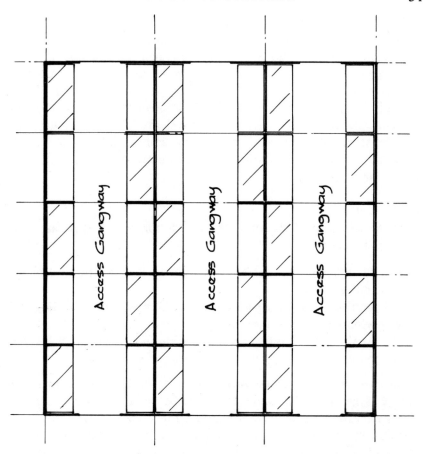

Fig 1.15   PLAN OF WAREHOUSE SHOWING
RACKING BANKED INTO 5m BAYS WITH
GANGWAYS FOR FORK-LIFT TRUCKS

located therein without affecting the efficiency of use while reducing roof span requirements dramatically. A column grid of 8 x 5 m would therefore provide a more economic solution. Secondly, there is no particular value in installing a brickwork perimeter to the building if an acceptable envelope can be provided by means of vinyl-coated, steel-sheet cladding. This is quicker to construct, ties in easily with the concept of a framework, and obviates the need for strip foundations around the entire perimeter of the building.

Continuing the considerations of the form that the framework should take, we should also look at the choice of materials. In situ reinforced concrete can be discounted as it offers no particular advantages. Indeed, it has some drawbacks insofar as it requires a high degree of quality control on mixing, placing and curing. Precast concrete might be considered but is somewhat heavy in the roof elements and could pose some fixing problems. The obvious material, then, is structural steel since it can be erected quickly, is easy to cross-brace and does not require fire protection when used for a single-storey structure.

Roof structure
As far as selection of the roof structure is concerned, there are three obvious choices (see fig. 1.16).

(a)   flat roofing with dome lights incorporating ventilation louvres in the cill sections;
(b)   monitor roof construction with the upstands providing top-hung windows for ventilation;
(c)   north-light trusses having opening windows at the valley section and fixed patent glazing within the sheeted area.

Flat roofing could cause problems with drainage falls since it is quite a large roof plan area. The decking might also be susceptible to leakage problems, which could affect the stored components.

Monitor construction is also primarily of flat roof format, but at two levels, and poses some of the same problems of roof drainage. Both flat roof and monitor roof forms offer advantages in minimising the volume of the structure below them in terms of space heating. In other circumstances this might offset the drainage difficulties in weighing merits against demerits.

In this case, however, space heating is not a consideration and the north light offers some benefits in that:

(i)   the trusses will be formed of angle sections and will use less steel than the flat roof or monitors;
(ii)   the pitched roofing provides better falls so that rain penetration through the cladding is less likely to occur;
(iii)   drainage can be via valley gutters falling to the ends of the building.

The final choice, therefore, is for a north-light configuration supported by steel columns, which are cross-braced for lateral

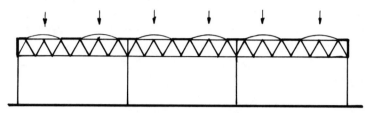

Flat roof - dome lights
most effective for direct overhead

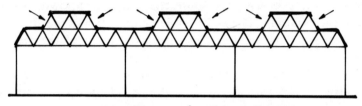

Monitor - raised sections
most effective for early and late

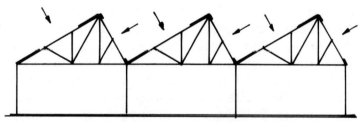

North light - trussed roof
effective for direct, early and late

Fig 1.16 ALTERNATIVE ROOF FORMS
SHOWING MAIN SOURCE OF NATURAL
LIGHTING AND ITS EFFECTIVENESS

stability. As mentioned earlier, the vertical cladding is coated steel sheeting, which offers good resistance to impact damage. The roof cladding, however, is not subject to the same risk of damage. This would utilise asbestos cement sheeting, which is cheaper to install and requires little by way of maintenance, there being no need to check the surface for possible corrosion. For details of roof construction see fig. 1.17.

Constructional techniques

The construction is split into phases and involves a specialist subcontractor for the erection of the steel superstructure, this being part of a 'design, supply, erect' package. The main contractor, however, is responsible for preparing the foundations and groundworks and for cladding the building. The only service requirements are for lighting and drainage and these, too, fall within the remit of the main contractor.

Access to the building, by fork-lift trucks, is at the same level as the external paved area. Because of this, it is logical to construct the ground slab prior to the steelwork erection, the weatherproofing of the external envelope at ground level being completed after cladding the framework. The procedures would therefore be to construct the foundations, install the surface-water drainage system, provide for electrical supply and then form the ground slab. At this point the steelwork subcontractor can be called in to erect the frame.

The building can then be clad, drainage down-pipes fitted, lighting installed and door gear put into place. The window gearing can then be fixed. The racking for the pallets can be installed next. This is bolted into the floor slab and tied back to the internal columns.

[Note: Both for the foundation piers to the framework and the floor slab to the racking it is essential that the level of the groundwork be consistent.]

Although the three case studies analysed are for very different forms of construction, there are some procedures that are common to these and other forms:

(1) *Taking possession of the site.* The main contractor, when taking over the site, has to assure himself of certain facts. Are the controlling dimensions, boundaries and topography in accordance with the contract drawings? If not, the development could be affected, both from the consideration of building sizes and their location and in the layout of the drainage system. Has a proper site investigation been

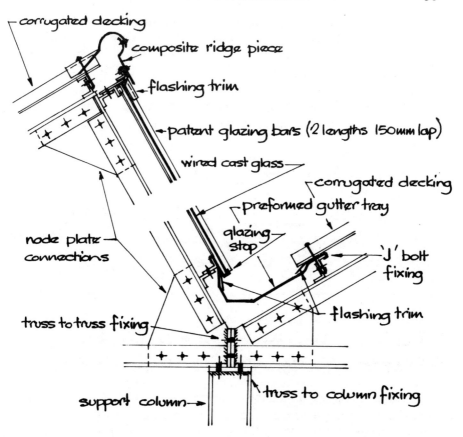

corrugated decking

composite ridge piece

flashing trim

patent glazing bars (2 lengths 150mm lap)

wired cast glass

corrugated decking

preformed gutter tray

node plate
connections

glazing
stop

'J' bolt
fixing

truss to truss fixing

flashing trim

support column →

truss to column fixing

Fig 1.17 NORTH LIGHT ROOF – DETAILS AT RIDGE & VALLEY SHOWING GLAZING, ROOF DECKING, GUTTERING AND ASSEMBLY

carried out? If not, it is the contractor's responsibility to ascertain soil-bearing characteristics and ground-water level, since these could affect the foundation sizes and excavation work. Where can waste excavation be disposed? If this is problematic, it may be that spoil can be used in the landscaping of the site. What services run across the site? Some might be utilised, some may need to be diverted and some might have to be closed.

(2) *Establishing levels and grids.* The datum of the site has to be

established and from this the various formation levels graded, using mechanical excavator/dozers. Depending on the ground conditions, this equipment will be either wheeled or tracked (a crawler system of caterpillar form like that used on tanks). The latter form of equipment is suitable for poor or variable ground conditions, since it spreads its load more evenly. However, it is not road-going and has to be delivered to and collected from site by a low-loader lorry (see fig. 1.18).

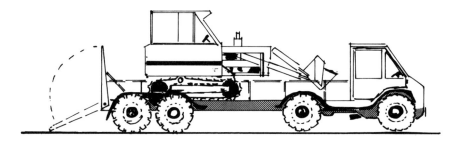

Fig 1.18 "Low-Loader" Lorry Carrying Caterpillar Tracked "Dozer"/Excavator

In some instances it might be useful to 'blind' the various areas immediate to the proposed construction so that the grids can be set out accurately. [*Note*: The grids referred to are constructional, being related to the positions of walls or columns, and should not be confused with survey grids.] Depending on the nature of the contract, a site square, for simple buildings, or a theodolite, for more complex, framed structures, will be used. In either case, some form of physical reference points need to be established to 'tie in' the grid location.

(3) *Ensuring safe working conditions*. All machinery and equipment used should be serviced and checked at regular intervals to minimise the risk of failure. This includes vehicles, plant, hand tools, scaffolding and temporary support works such as trench strutting and shuttering to in situ concrete. Any procedures used should conform with manufacturer's instructions, construction regulations or the contract specification, whichever is appropriate. All materials and components should be stored safely so as to eliminate any risk of breakage or collapse. Also the tidiness aspect should be drilled into all members of the work-force. All scrap materials and components should be put in skips to avoid the risk of accidents. Although not 100% true, the

statement 'A tidy site is a safe site' does have a considerable degree of logic in terms of accident prevention. It is also necessary to ensure that adequate temporary lighting is provided, particularly in the winter months, to produce good working conditions.

(4) *Site security*. Perimeter fences and occasionally hoardings, gantries and 'fans' should make the site secure from trespass and from pilferage of goods. Hoardings and fans should protect the general public or adjacent property from dirt and debris. On some sites, guard dogs and/or sophisticated security systems are necessary.

(5) *Accommodation*. On all but the smallest of sites, some level of hutting must be provided. This will generally include toilet facilities, changing accommodation and a sick-room (including first-aid equipment), together with offices for personnel such as the site agent and/or site foreman, resident engineer or clerk of works, and a room for site meetings. On large sites, canteen facilities might also be provided but the scope of such hutting will vary according to the size of contract. Caravans, portable cabins and demountable buildings are readily available for hire and many builders prefer to use a leasing system rather than buy their own units. Obviously, it is necessary to establish a water supply, sanitation and drainage, temporary power and lighting, and telephone communications to service the site accommodation.

Assignment 1.1
Produce a reasoned argument for adopting an alternative form of structure to the one chosen in case study 1.2, expanding your reasoning with well-annotated sketches of pertinent constructional details. Some background reading of current technical journals may be helpful in answering the assignment.

Assignment 1.2
Develop a monitor roof system for the warehouse covered in case study 1.3 and compare the approximate weights of steelwork used. Also cost the difference in the forms of roof deck/cladding used in each instance. A current schedule of material costs will be necessary to answer this assignment.

2 Internal Services

In Volume 1 we discussed briefly the carcassing out of a building for service requirements. In this chapter we will take a more detailed look at the provision for, and installation of, internal services as applied to domestic and commercial buildings. We tend to accept services of most kinds as commonplace in the buildings being constructed today in the U.K. but it is worth noting that it was not always the case. Even in the 1950s some houses did not have electricity for lighting and power and this was not confined just to outlying villages but was also true of some buildings in and around London. Water supply in the early part of the 20th century was only available from pumps and stand-pipes, and gas supplies are still not always provided for new building work. Telephones are more usual than was the case as late as the mid 1960s, and closed-circuit television for commercial buildings is still thought of as somewhat exotic. Let us look at each of the services in turn as applied to particular situations, since not all situations can be covered in the space of one chapter.

COLD WATER SUPPLY

The service from the main is controlled via a stopcock sited close to the boundary of the property. There is a further control valve just inside the building, which can stop the supply. These two controls are used to isolate the supply for different reasons. The *stopcock* can be operated by the water authority to stop the supply in the event of:

(a) a burst in the main;
(b) a burst between the stopcock and the property;
(c) a leakage causing contamination of the supply;
(d) non-payment of water rates or metered supply.

The *stopvalve* can be operated by the tenant or owner of the building to stop the supply in the event of:

(a) a burst or leakage within the system;

(b) extension or adaption of the system, e.g. an additional hand-basin;

(c) for refurbishment of the system, e.g. a new water storage tank;

(d) vacating the building for an extended period, such as a holiday, as a precautionary measure.

The water supply inside the building is then routed to provide a service to the user. At least one pipe is taken off this 'rising main' to provide drinking water. In the majority of cases a storage supply is required and the main supply rises through the building to a cold-water storage tank at a high point (hence the name). Some buildings, however, do not have this facility. This form of supply is called 'direct mains', with all services being tapped off the rising main, which feeds a much smaller storage tank for supply to the hot-water cylinder only. This reduces the chance of frozen pipework but does make the supply subject to mains pressure variation and possible water hammer.

For this case let us consider a detached house of high specification to include two bathrooms, one with shower and bidet, a ground-floor toilet and wash-basin and a utility room.

The number of fittings is higher than that usually encountered, and to cope with this a standby water storage tank is provided using an 'in-tandem' system. Another innovation is the inclusion of a water softener in the supply, situated fairly near the stopvalve. Immediately above the stopvalve is a drain tap so that the complete system may be drained down if required. Because the laundry room is adjacent to the kitchen, a feed from the rising main serves both sinks with drinking water.

The first of the two storage tanks is known as a cistern since it incorporates a ball valve and overflow pipe to replenish the water used. It also has a feed pipe into the second storage tank, which feeds the cold water supply to the sanitary fittings, the hot-water cylinder and the 'header tank' to the central heating system. Each of these supplies is also fitted with a stopvalve so that any of them can be isolated without affecting the other two. This might occur when draining down the central heating system for the addition of a radiator (if extending the property) or replacing the hot-water cylinder when it is faulty. [Note: The header tank to the central heating system is unlikely to operate on a regular basis.]

An isometric in line form (fig. 2.1) shows the cold water supply. Other aspects of the overall water system – hot water supply and central heating – will utilise the same components but will each be shown separately to aid identification.

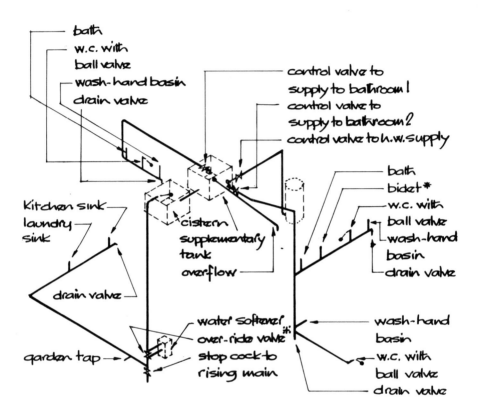

bath
w.c. with ball valve
wash-hand basin
drain valve

control valve to supply to bathroom 1
control valve to supply to bathroom 2
control valve to h.w. supply

bath
bidet *
w.c. with ball valve
wash-hand basin
drain valve

Kitchen sink
laundry sink

cistern
supplementary tank
overflow

drain valve

garden tap →

water softener
over-ride valve *
stop cock to rising main

wash-hand basin
w.c. with ball valve
drain valve

Fig 2.1 COLD WATER SUPPLY

[Notes: * bidet used has taps, rim-feed would require separate supply pipe
*: over-ride valve for use when water softener is recharging or being serviced]

HOT WATER SUPPLY

This is also available as a direct or indirect system, the latter being far more common since it has a stored supply contained within a hot-water cylinder. The direct system does, however, possess some advantages since it provides 'continuous' hot water within a limited capacity. The most common direct system is the 'multi-point', which feeds several sanitary appliances. The cold water feeds through a microbore tube coiled round a gas burner, which ignites when water is drawn off. A large surface area is heated and hot water is drawn off at the required appliance. If more than one appliance is used at the same time, however, demand may outstrip supply with a subsequent drop in temperature. Where such demand seems likely, the multi-point may be supplemented by an individual, small, sink water-heater.

Stored hot water can be heated by several methods:

(a) *Back boiler from an open fire* — this heats the water in the cylinder directly.

(b) *Electric immersion heater* — a single or twin element heats the water in the cylinder directly. In the case of the twin element, a short element heats only the upper part of the cylinder (this being adequate for normal usage), the second element being used only for bulk heating.

(c) *Gas circulatory system* — the water feeds through a micro-bore, similar to that of the multi-point boiler, and is circulated back into the cylinder until the required temperature is achieved. This system has two feeds for heating part or all of the cylinder. This is again a direct heating technique. [*Note*: both the multi-point and the circulatory boiler require flue terminals.]

(d) *An indirect heating coil or jacket* inside which water from the central heating system is passed. This coil (or jacket) then heats the water in the cylinder until the required temperature is reached.

(e) *Solar panels* — water passes through panels located in a position to trap the optimum amount of sunlight and is heated and stored. This system is generally treated as secondary heating and is often supplemented by one of the other techniques.

Most central heating systems have the facility to heat the domestic hot water supply. The exceptions are electric storage heaters (whether independent, fabric or warm air) and some gas warm-air systems. A back-up system of an immersion heater is, however, usually installed in the event of the boiler breaking down, or for summer usage if the boiler is located in a position where it emits too much heat. When the immersion heater is the only source of hot water, it is logical to use a very large capacity cylinder with a high level of insulation — sprayed on rather than in quilted jacket form — using off-peak electricity. In these situations it is sensible to use an electric shower heater unit, which can heat the shower water directly at the required temperature independently of the main hot water supply. Figure 2.2 shows the hot water supply that complements the cold water supply illustrated in fig. 2.1.

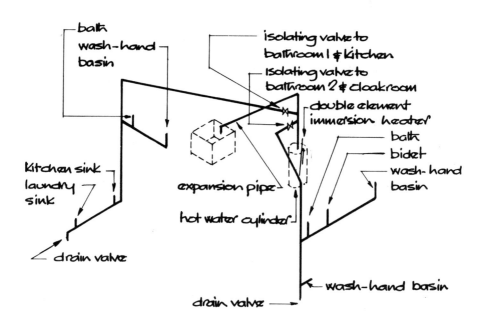

Fig 2.2 HOT WATER SUPPLY

CENTRAL HEATING SYSTEMS

Central heating, as opposed to unit heating, falls into three basic categories, although these can be subdivided. These are:

(a) *Radiators* – either in panel form, skirting runs, fan-assisted convectors or a mixture of two or three forms. These can use
(i) large-bore pipework (only for gravity feed systems with no pump),
(ii) small-bore pipework (the most usual form), using either one-pipe or two-pipe circulation,

(iii) mini-bore pipework, using two-pipe circulation only.

(b) *Ducted warm air*, using either vertical ducting with ventilator grilles discharging directly, or a mixture of horizontal and vertical ducting. The latter requires a suspended floor at ground level and ceiling-level grilles at the upper floor, the ducting being located in the roof space. A larger capacity air-handling fan is also required for this method than is required for vertical ducting. [*Note*: The efficiency of ducted warm air systems can be improved by insulating the duct runs and fitting controllable shutters on the ventilator grilles.] The air supply for warm-air heating can be either direct, using fresh air, or recirculatory, the latter requiring extract grilles at points on the system.

(c) *Fabric heating*, using either the floor, the ceiling or the walls (the last rarely being used). The underfloor heating gives a very low output but has disadvantages where the movement of occupants is slight, e.g. in old people's homes. The ceiling heating is to some extent radiant heating and can cause headaches. Wall heating means that the use of wall furniture or picture hanging is restricted.

Returning to the house used to illustrate the hot and cold water supplies, we will use a radiator system of heating but with a high specification. Rather than use the conventional room thermostat, we shall fit thermostatic valves to each radiator in the system. Also, the two floors will have separate flow and return circuits with control valves to allow one or both circuits to operate. A programmer controls the overall operation of the system, providing for time switching (i.e. continuous, once on and off, or twice on and off) and for selection (i.e. hot water only, hot water and radiators on one or both cycles). Additional to the radiator valves the boiler has a thermostat control and a motorised valve is located in the indirect hot-water cylinders feed. Figure 2.3 shows the central heating runs only as associated with the hot and cold water supplies illustrated in figs 2.2 and 2.1.

The fuel supply should not be confused with the type of central heating used, since the three varieties described are not limited to one fuel. Radiator boilers can use gas, oil or solid fuel (anthracite or coalite). Warm-air ducts can be served by an off-peak electricity store, or by gas-fired or oil-fired heat exchangers. Fabric heating can utilise electrical elements or be formed by micro-bore piping runs served by gas- or oil-fired boilers. It might also utilise solar heating but this would demand a huge volume of stored water.

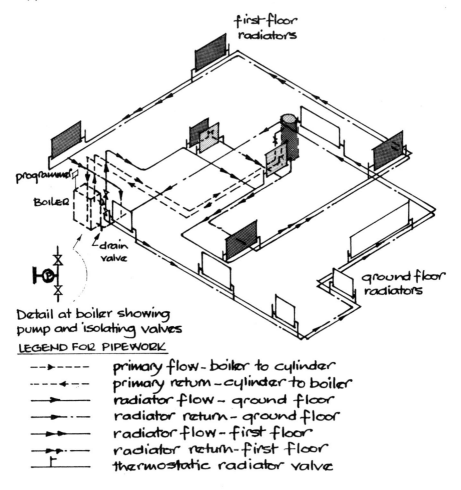

first floor radiators

programmer

BOILER

drain valve

Detail at boiler showing pump and isolating valves

ground floor radiators

LEGEND FOR PIPEWORK

--►----	primary flow - boiler to cylinder
----◄--	primary return - cylinder to boiler
──►──	radiator flow - ground floor
──►-·-	radiator return - ground floor
──►►──	radiator flow - first floor
──►►-·-	radiator return - first floor
──┌──	thermostatic radiator valve

Fig 2.3 CENTRAL HEATING SYSTEM

GAS SUPPLY

As described in Volume 1, the supply to the house is installed by the service authority to a gas meter located inside the property. A governor/filter is fitted to regulate the pressure, this being particularly important when natural gas (North Sea supply) is used. A control valve (gascock) is fitted to the supply pipe feeding into the meter so that it can be closed off in the case of an emergency or if the house is left unoccupied for a while. The carcassing out is of a radial form, the pipe being 'spurred' for the required fittings. Where a gas supply is fitted it is customary to provide for the cooker, the central heating boiler and possibly a gas fire. Alternatively, if unit heating by gas-fired convectors is intended, each unit will be supplied. A control valve is supplied to each point used and where the point is provided only for possible future use the supply is capped. The piping is usually of large diameter mild steel gas barrel but can be reduced in diameter and chromed externally for improved appearance if serving a gas fire or gas poker to an open solid-fuel fire. A flexible hose connection can be fitted at the cooker service pipe to allow movement for cleaning. [*Note*: The gas meter should be located in a ventilated position, usually the garage, to avoid excessive heat, which might produce an explosion.] A gas supply is shown for the house considered in the hot and cold water supply examples. (See fig. 2.4.)

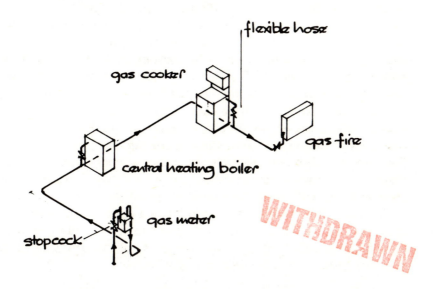

Fig 2.4 DOMESTIC GAS SUPPLY

ELECTRICITY SUPPLY

The main electricity distribution network servicing a district is 415 V three-phase supply and from this each domestic dwelling is provided with a 240 V single-phase supply, this being adequate for normal domestic usage. In the same way as with the gas supply, the electricity is fed through a meter. However, whereas the gas supply has a control valve to the authority's side of the meter, the electricity supply has a sealed service fuse box in this location. If the installation is to a new building, there should be no problem since the service fuse should be more than adequate for the load likely to be used, being based on the number of outlets and fittings. When the system is upgraded, however, it is advisable to ask the authority to check the rating of the existing service fuse. If this were to be overloaded, the complete supply would be cut off and could only be reinstated by the service authority. In some instances a separate supply is fed through a white meter, which allows off-peak electricity for use at certain periods only. This is used, as mentioned earlier, for storage heaters and bulk water heating, the unit cost being about half that of the conventional supply.

Let us consider the conventional supply in more detail. After passing through the meter the supply cables are fed into a consumer unit. This translates the single supply into a convenient number of circuits, some of which serve the lighting and others the power sockets. A radial supply is made to major items such as a cooker, which has a higher fuse rating. The consumer unit has a main control switch, which can shut off the entire system, and a series of fuses (or mini-circuit breakers), which control each circuit. Thus if the circuit overloads or develops a fault, the fuse *for that circuit only* blows (or the circuit is broken) without the other supplies being affected. Figure 2.5 shows a simple layout of service fuse box, meter and consumer unit with alternative fuses or circuit breakers.

The main difference between a fuse and a mini-circuit breaker is in their operation. When overloaded, the wire in the fuse melts and breaks whereas the circuit breaker 'trips' the circuit and can be reset. The fuse system has some disadvantages, namely:

(a) The fuse wire has to be replaced with the correct rating of wire — 5 amp for lighting, 20 amp for outlets, 30 amp for cooker and possibly immersion heater.

(b) It is not readily apparent which fuse has blown unless the effect has been sufficiently severe to scorch the fuse holder.

(c) In order to isolate the circuit the fuse holder has to be removed.

The circuit breaker, however, can be reset without the need for repair (remembering to switch off the main control of the unit before doing so) and indicates which circuit is at fault by tripping out the setting

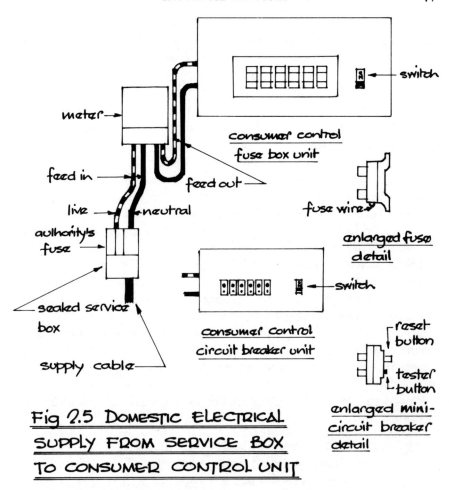

meter

feed in

feed out

live — neutral

authority's fuse

sealed service box

supply cable

consumer control fuse box unit

switch

fuse wire

enlarged fuse detail

consumer control circuit breaker unit

switch

reset button

tester button

enlarged mini-circuit breaker detail

Fig 2.5 Domestic Electrical Supply from Service Box to Consumer Control Unit

button. This can also be done by using the trip button if you wish to isolate the circuit. The only disadvantage of this system is its cost in comparison with the fused system, but this may be offset by its convenience of operation.

The carcassing out is best done during the construction stage while floor joists are exposed, since it is easy to thread the P.V.C.-clad cable through holes drilled at the middle of the joists to avoid nailing through the supply. In timber-framed construction this is equally true of the vertical studding but in traditional construction it is necessary to 'chase' the blockwork so that cables for wall sockets or light switches can be recessed.

There is no actual limit to the number of outlets that can be accommodated on a circuit. Logically, however, it is sensible to separate the outlets by a series of circuits so that not too many are affected in the event of a fuse or break. For a small two-storey house this might result in two lighting circuits (one per floor) and two ring circuits (one per floor) plus a radial supply to the cooker. For the house described in the example considered for the hot and cold water supply, this degree of separation might be thought inadequate. The lighting might therefore be further divided to provide a separate circuit for the garage and kitchen/utility area (lighting in the area of the supply meter), and possibly two circuits on the first floor to include the roof storage space. The ring circuits for power outlets would have a separate supply for the central heating system, possibly a supply for the kitchen/utility area alone (washing machine, tumble dryer, etc. rated at 30 amp) and perhaps a radial supply to the garage for use of power tools (also rated at 30 amp).

The principle of the ring circuit is that the cable runs around the socket layout system, with each socket feeding directly into the circuit. In practice it may be uneconomic to run the full perimeter of the house and it is possible to spur off the circuit to feed awkwardly located sockets or fittings. [*Note*: Recommendations are that a ring circuit should not supply more than 100 m² of floor area and that the number of spurs does not exceed half the total number of sockets on the circuit.]

Power socket outlets are available to a variety of specifications including:

(a)   single three-pin socket panel   unswitched;
(b)   double three-pin socket panel − unswitched;
(c)   single three-pin socket panel − switched;
(d)   double three-pin socket panel − switched;
(e)   fused connection units − unswitched;
(f)   fused connection units − switched with neon;
(g)   shaver socket − unswitched;
(h)   shaver socket − switched;

[*Note*: (g) and (h) are designed specifically for use in the bathroom and are two-pinned.]

(i)   composite cooker and three-pin socket − switched with neon.

The advantage of using neon indicators for fused connection units and cooker panels is that they can be checked from a distance to see if the appliance is on or not, a typical benefit being an immersion heater switch located on the landing at the top of a stair flight.

Because the labour content of wiring and connecting a double socket panel is no more than that of a single panel, and since the component cost is relatively small in the overall costing of the circuit, it makes good sense to provide double rather than single socket panels when rewiring or carcassing a supply. It is also worth installing switches since they provide added security even though each individual plug is fused. The layout for the house described earlier in the chapter is shown in fig. 2.6.

Safety aspects that need to be considered in the wiring up of the electrical system are:

(a) the system must be fully earthed;

(b) the cables should be protected from accidental damage as far as is practicable;

(c) light switches to areas such as kitchen, laundry and bathroom should be pull-cord operated;

(d) no socket outlets other than that for an electric shaver should be located in the bathroom;

(e) the correct capacity fuses should be fitted to all fixed appliances;

(f) sockets should not be located where they might be liable to impact damage;

(g) cable runs should not be covered by thermal insulation quilts, since this may promote overheating and the risk of electrical fire (the most likely area of risk is the lighting runs in the roof space);

(h) the finished installation must be checked for leakage;

(i) all components must meet B.S.S. requirements.

The consideration in this chapter has been pitched at a low-rise domestic building level. Larger buildings may require distribution boards at various locations to split the supplies. This is dealt with more fully at higher certificate level.

It should be realised that for both the electrical and the gas supplies the meters must be readily accessible for reading by the service authorities. It is sensible, then, to locate them with this in mind. The garage is a useful place since access can be provided without the need for entry to the house. Alternatively, they may be sited in a cupboard located on an outside wall, with a window allowing them to be read easily. The traditional 'under the stairs' location is probably one of the least convenient positions for ease of access.

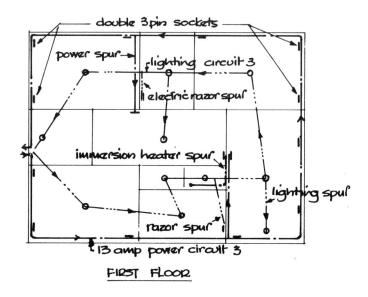

FIRST FLOOR

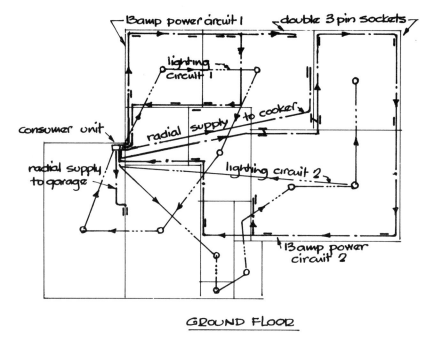

GROUND FLOOR

Fig 2.6 POWER AND LIGHTING CIRCUITS (SWITCH RUNS NOT SHOWN) TO DOMESTIC PROPERTY

SANITATION AND DRAINAGE

In order to remove soil- and waste water from a building, a system of internal drainage pipes feeding into a main stack is necessary. Regulations concerning the location of such pipework are subject to change, the requirements at one stage being that *all* pipework should be internal. Current requirements have moved back in some respects to those operable some 40 years ago and for housing up to three-storey height the pipework can be located on the outside of the wall. Where it differs from the early provisions is that waste water cannot discharge into a 'hopper' (an open catchment point) whereas this was a common feature of housing built up to around 1945–50.

There are two common systems of sanitation and drainage in current use:

(a) single-stack, where all fittings discharge into one vertical stack pipe;

(b) two-stack, where the toilets feed into one vertical stack pipe (soil stack) and the other fittings feed into a separate stack.

The single-stack system is a direct result of the analysis of existing systems, and how they might be improved, by the Building Research Establishment. It has several criteria that need to be observed, including pipe sizes, lengths of waste runs, falls, restricted areas of entry to the stack and depths of water trap seals. These criteria are spelt out in the appropriate B.R.E. Digest on sanitation and drainage.

The two-stack system was the 'quality' method used in preference to the early single-pipe system. It was not subject to the risks of break of water seal that might occur in the combined system of waste and soil. Unfortunately, it was also rather more expensive to install. It was this factor that promoted the research culminating in the development of the single-stack system. With modification this can be used for medium- and high-rise construction. Because the number of appliances is greater the rate of discharge into and from the stack increases dramatically in comparison with that for a single dwelling. There is then the resultant risk of siphonage, which could draw water from the traps of individual appliances and break the trap seals. Anti-siphonage piping is, therefore, built into the system.

Whichever system is adopted, it would be illogical to feed waste from the ground-floor fittings into the vertical stack pipe as this would be uneconomic. The soil-water from the W.C. feeds directly to the inspection chamber, and the waste water feeds into a back inlet gully. This may be closed or grilled but in the case of the grilled gully the discharge pipe must finish below the level of the grille.

The materials available for use are cast iron, P.V.C., galvanised steel, pitch fibre and copper. The last of these is rather expensive and not really a very sensible choice to make. Cast iron and galvanised mild steel are rather heavy to handle and support, and pitch fibre is considered by many people to be of a low specification. The most popular material is therefore P.V.C., since it is light to handle and support and also very simple to joint, all work being dry jointing incorporating rubber seal rings. Figure 2.7 shows the sanitation pipe-work for the fittings in the example described earlier in the chapter.

[*Note*: Whichever system is used, it must be ventilated using a fresh air inlet in the final inspection chamber and a vent pipe at the highest point on the system.]

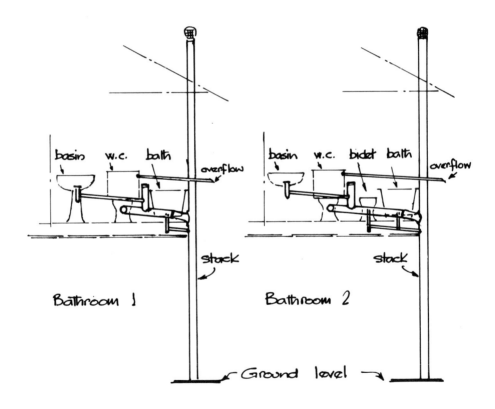

Fig 2.7   SANITATION TO UPPER FLOOR

COMMUNICATION SYSTEMS

Domestic construction is concerned almost exclusively with telephone communications, and since, in the U.K., all work is controlled by British Telecom, the service installation is of little interest to the builder. The supply is either by underground or overhead cable, which is tapped off and metered for one or more 'external' telephones, possibly with extensions to the upper floor. All internal wiring is fitted by the service authority, the telephone(s) being on rental, and the wiring is of such small diameter as to be unobtrusive if surface mounted.

It is in commercial construction that the communication systems are more sophisticated. The telephone layout may be split into 'external' and 'internal' services but there is little value in this since the number of receivers is double that of a combined service. By using a switchboard it is possible to accommodate a large number of extension lines on one external telephone. Many offices, however, have a small bank of external telephones so that the system does not become overloaded. Modern electronic control systems allow for programming of the individual extension telephones. This can include automatic re-routing, transfer of call, open conversation (on more than one extension for an outside call), automatic search and memory bank of regular numbers in use. Obviously the wiring is more complex for such systems and boxed metal ducting is used. This may be in peripheral runs, located within pressed metal skirtings, or radial, located in floor slabs, floor screeds or suspended ceilings. The floor locations provide for a pre-determined number of terminals, since the routing of the ductwork cannot easily be modified. However, the ceiling location gives some additional flexibility although additional spurs might require drilling through the floor slab for work above, or adaptation of the partition work if routed below. The logical answer is to be generous in estimating the likely number of terminals, since no harm is done if they are not all used.

An additional communication system used in offices is computer terminals for visual display units (V.D.U.s) where centrally stored information can be called up on to a screen some distance from the data store. Cable runs can be located in a similar fashion to those for telephone extensions and can possibly share the ducting if space permits.

Closed circuit television (C.C.T.V.) is also used in commercial buildings for both communication and security services and although used in some offices it is more popular in shops and department stores. The cables are ducted as with V.D.U.s and there will generally be control consoles, housed behind access panels, that need to be accommodated. Vertical service ducts may be of sufficient size to warrant specific provision but these should be located in positions secure from easy outside access.

Supermarkets, warehouses and factories utilise a public address system, which is really a one-way aural communications system. It is generally housed within the suspended ceiling area, or possibly in conduits for warehouses, and requires no special support.

HEATING, VENTILATION AND AIR-CONDITIONING (H.V.A.C.)

The most usual form of installation is that operating from a central source utilising pressed metal ducting. Requirements for provision of the system are quite substantial. They involve one or more plant rooms, to house the compressors, large vertical ducts and horizontal ducting, either at the perimeter of the building or, more usually, within the suspended ceiling. In the latter case this might involve providing holes through downstand support beams, which must be allowed for at the structural design stage. Hanger supports are necessary for the horizontal ducts and provision can be made by casting dovetailed slots in the concrete flooring at its underside.

Unit air-conditioning overcomes many of these problems since each unit contains its own filter, refrigeration and heating components. However, these units must be sited on external walls so that they can be ducted directly to the external environment for fresh-air intake and stale-air outlet vents. Provision for this must be made in the vertical cladding to the building but this is minor compared with the allocation of a plant room and suspended ceilings, neither of which are needed. Balance between units can be on an electronic sensory system to allow for compensation of solar gain, orientation (varying exposure conditions) and heat-generating office appliances. The wiring runs can be located in the same small ducting system used for telephones and V.D.U.s using colour-coded separating clips.

FIRE BREAKS

The vertical ducting housing sanitation and drainage (possibly hot and cold water supplies) C.C.T.V. and H.V.A.C. systems is quite substantial. It may be located to the rear of the lift shafts or in a sandwich between washrooms and general offices and can also house dry-riser fire-fighting pipes. It does, however, form a flue through the building, which in the event of a fire could act as a ready distribution system. To avoid this, it is *essential* that a fire break is formed at each floor. After the various pipes, conduits and pressed metal ducts have been fixed, a surround of expanded metal lathing should be fitted across the remaining voids in two layers. The lathing acts as a permanent shutter and can support a lean mix of concrete, the underside being rendered. This concrete layer is built up to a thickness of approximately 100 mm, which forms an effective fire barrier.

The H.V.A.C. ducts themselves can also act as fire flues in both the vertical and horizontal planes and must be fitted with fire shutters, which close the ducting at certain temperatures. These are also fire breaks but are intermittent rather than permanent so that the system can function under normal conditions.

FIRE-FIGHTING AND ALARM SYSTEMS

Because commercial and industrial buildings are more densely populated and contain more equipment and data storage than domestic buildings, it is a statutory requirement to protect both the occupants and the equipment in the event of a fire. As already mentioned, 'fire risers' are built into the structure so that they can be quickly connected by hose reels and the water supply operated to control and fight a fire before it has become fully established. Warning can be given by smoke sensor alarms, which alert the people in the building to the possible outbreak of a fire. Supplementary fire alarms can warn the remaining occupants some distance away from the source to allow controlled evacuation of the building.

Some areas may be particularly prone to damage in the event of fire and in such situations a sprinkler system, located at ceiling level, may be installed. The pipework may be cast in, fixed directly or suspended on strapped hangers from the floor slab. The sprinkler roses contain heat-sensitive valves. One valve can set off the whole system within a specified grid and unfortunately the sprinklers can be set off by irresponsible jokers using lighted matches.

In Volume 1 we talked about means of escape, this being linked with the occurrence of fire but not strictly part of the fire-fighting, i.e. it allows for controlled evacuation but does not fight the fire. However, if a fire is contained, it will exhaust its oxygen supply.

Fire doors are therefore constructed to withstand combustion and/or distortion for specified periods. So that fire does not pass around the frame or lining of the door, or fresh oxygen pass through to feed the fire, an automatic heat-generated sealing is provided. Intumescent strips are built into the door and, having a high coefficient of expansion, seal the gaps between the door and its jambs, head and threshold on reaching the appropriate temperature.

LIFTS AND ESCALATORS

These are commonplace in commercial buildings and will be dealt with more fully at higher technician study level, as indeed will the other services covered in this chapter. Suffice it to say here that they are major items of service equipment. They must therefore be catered for at the initial design stage of the structure, both for the floor openings (which would cause severe structural problems if made to an existing building) and for the location of associated plant. A typical lift shafts/motor room layout was discussed in case study 1.2 in chapter 1.

Assignment 2.1

Produce an isometric layout of the hot and cold water supply for a typical semi-detached house of the type illustrated in fig. 5.2 of Volume 1. Indicate the position of all valves and controls, assuming that the hot water supply is direct via a back boiler to an open fire. Describe the difference between compression and capillary fittings for jointing the pipe runs and assess the merits of each. The use of manufacturers' catalogues will be helpful in answering this part of the assignment.

Assignment 2.2

Using the house layout illustrated in fig. 1.12 of Volume 1, consider a radiator central heating system using two circuits (one for each floor). The control is by a room thermostat for each circuit. Explain how this operates and suggest logical sitings for each thermostat, giving reasons for your suggestions. Depending on where you have located them, indicate the setting at which 'cut-out' should occur, giving a reasoned explanation.

Some additional reading would be helpful in answering this assignment.

Assignment 2.3
Using the layout for a three-storey construction of flats and maisonettes illustrated in fig. 5.8 of Volume 1, produce an isometric layout of a single-stack sanitation and drainage system to meet the requirements of the current Building Regulations.

Reference to the relevant B.R.E. Digest will be useful in answering this assignment.

3 Finishes: Internal and External

Finishes, as they affect domestic construction, were dealt with in some detail in Volume 1, chapter 4. The principles relating to their purpose, behaviour and deterioration are much the same for commercial and industrial construction. In this chapter we will concern ourselves primarily with finishes in relation to commercial construction, this being the area where the widest variety occurs.

INTERNAL FINISHES – CEILINGS

These will depend to some extent on the materials used for the roof or floor slab. A timber flat or pitched roof will probably use plasterboard whereas a steel or concrete roof will not. Also, if a suspended ceiling is used, the finishes may vary depending on their required function. Let us consider some examples of finishes used to best advantage.

Example 3.1
An office building of flat slab construction, similar to case study 1.2, utilises the concrete soffit as a self finish ready for decoration. How can this be achieved?

First the shuttering must be to a consistent level across the entire wing of the building. By using 'table-form' shutters, panels of 2400 x 1200 mm ply decking can be braced sufficiently to avoid sag across its span. Each table form is on a trestle, which can be adjusted for level at the four supports. Thus, by careful alignment, all the table forms can be levelled accurately.

The quality of the shutter face must be capable of producing a smooth finish. It is possible to produce plywood with a film of plastic to one face and this forms the shutter. The joints are taped and foam strips rammed home from below to stiffen the joint line. The concrete is vibrated sufficiently to eliminate air voids at the shutter face so that when it is struck the surface should be free from major blemishes. Slight projections at the joint can be sanded away and small imperfections in the surface filled with a fine textured paste. The surface is now ready for painting. [*Note*: If a stippled decorative finish is adopted, there is no need to fill the minor imperfections since these will be masked by the decorative stipple.]

Example 3.2
An office building uses suspended ceilings to house services and requires some damping of typewriter sound to be provided. How can the ceiling finish cater for this?

The suspended ceiling needs to be supported and must provide access to the services. A grid of inverted T sections on hangers will allow for panels of three types:

(a) those housing ventilator grilles for the air-conditioning;
(b) those providing some sound-absorbent properties, i.e. fissured surface acoustic tiling (this can also be used for (a));
(c) diffusers of translucent plastic sheeting for the light fittings.

By stripping the T housings with felt or neoprene, or possibly nylon pile, each panel can be 'damped' against vibration. The combination of an absorbent surface and a damped support grid will reduce noise and provide the functional elements of lighting and access as specified.

Example 3.3
A jewellery showroom requires a fully illuminated ceiling but must have sufficient ventilation to extract cigar smoke over its general area rather than at specific duct grilles. How can this be done?

As with example 3.2, a grid of inverted T sections is provided. However, the ventilator grilles are not located at ceiling level but at a higher plane, and they are so shaped as to spread over a large collection area. The ceiling panels are of rigid plastic formed in an 'open waffle' pattern and are white to give maximum reflective quality. The light source is positioned above them in sufficient numbers of lamps to produce an even spread over the whole area.

Figure 3.1 shows details of the solution to examples 3.2 and 3.3.

INTERNAL FINISHES – WALLS

These vary from the purely functional to the purely decorative, the

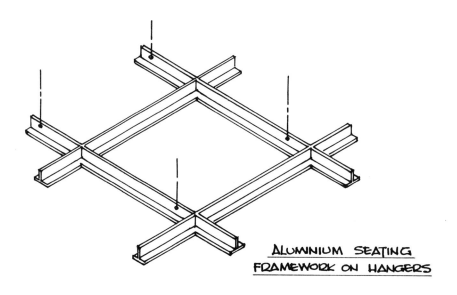

ALUMINIUM SEATING
FRAMEWORK ON HANGERS

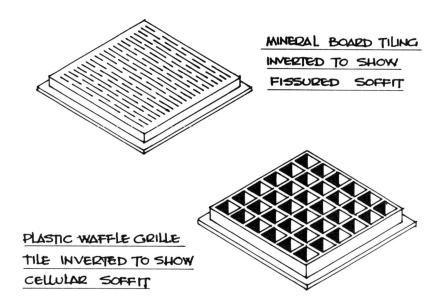

MINERAL BOARD TILING
INVERTED TO SHOW
FISSURED SOFFIT

PLASTIC WAFFLE GRILLE
TILE INVERTED TO SHOW
CELLULAR SOFFIT

Fig 3.1   SUSPENDED CEILING DETAILS
SHOWING ALTERNATIVE TILING TREATMENT

ideal finish incorporating elements of both characteristics. The finish relies to some extent on the 'background' or wall to which it is applied, and whether it is applied directly or indirectly. Let us examine some requirements for particular conditions and suggest how they might best be met.

Example 3.4

A director's office has one wall glazed from floor to ceiling. The flank walls are partition walls to other offices, with the corridor wall adaptable to any form. The flank walls must be sound absorbent and the corridor wall (non-structural) must give an air of opulence.

Since the flank walls are normal demountable partitions, it will be necessary to apply a finish that will relate to the other wall while offering acoustic qualities. A theme of natural materials is adopted throughout the office and the finish adopted for the flank walls is deep-fissured cork tiles bonded to the partitions. The corridor wall is of a short run and should be as impressive from the outside as it is from the inside. A system of deeply moulded hardwood sections jointed in the recess is selected, and the door is of the same construction so that it is concealed on the inner face (see fig. 3.2). The hard-

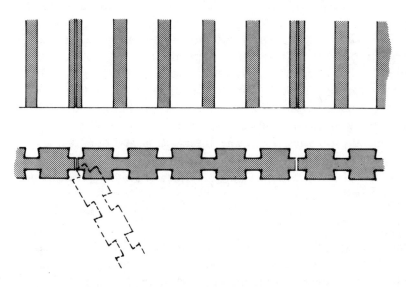

Fig 3.2 DETAIL OF SOLID HARDWOOD DOOR AND WALL PANEL SCULPTED TO MASK OPENING

wood is oiled to effect a warm appearance and the colour scheme of the cork and wood is echoed in the curtaining and carpets.

Example 3.5

A school corridor is of brick wall construction and is susceptible to impact forces. Children are to be discouraged from sliding along the walls but the finish must not look clinical.

In this case the solution is obvious: use the wall material to its best advantage. Select the type of brick carefully for use as a self finish. Do not use sand faced flettons because the surface will be defaced and the clay will show through, producing an unsightly appearance. Do not use sand lime or calcium silicate bricks because they are too regular and produce a 'regimented' appearance. To give a warmth, texture and durability, fired clay facing bricks should be used, with recessed joints in a coloured mortar that harmonises with the brickwork.

Example 3.6

A 'fast-food' restaurant has structural walls of reinforced concrete. The shuttering has produced a good regular finish but the walls will need to be amenable to easy regular cleaning to preserve a hygienic appearance.

Since the surface is regular it will not be necessary to provide a plaster background for tiling. Glazed ceramic tiles are adopted, with some picture tiles designed to relieve the monochromatic effect. The concrete background should first be treated with a bonding agent. Then, the tile adhesive can be combed on and the tiling set out from floor to ceiling working sideways from a vertical setting-out line located centrally on each wall. The intersection at floor level is of special curved quarry-tile skirting to form a continuous tiled surface. [*Note*: A 'ledge' is avoided so that no surface is available for dirt or grease to settle.]

Example 3.7

A hostel is to have all walls plastered and emulsioned so that redecoration can be carried out quickly and easily. What special precautions should be made in the selection and application of the plaster?

A dense gypsum plaster should be used as it is less prone to damage than lightweight plaster, for example by chairs, beds, etc. being pushed against the wall. Wherever openings occur in a wall, the variation of temperature is accelerated by air currents. Corners should therefore be reinforced to minimise thermal cracking, using expanded metal lathing. Similarly, external corners to walls at recesses, window reveals, etc. are liable to accidental chipping and should be prepared with expanded metal cornering (see fig. 3.3 for lathing and cornering details). Manufacturers' instructions with regard to drying times should be followed precisely to achieve the best results. All second fixings should be made while the plaster is still drying so that impact forces do not crack the finish.

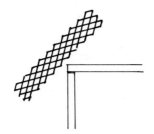

expanded metal lathing
pinned at corner of
opening prior to plastering

expanded metal
cornering pinned
to external wall
corner

Fig 3.3 EXPANDED METAL LATHS USED AS KEY AND FOR PROTECTION TO PLASTER

Obviously it is not possible to cover all contingencies in four examples but they do show the wide variations that occur. It is important to remember that walls provide the largest surface area in a room. Therefore, if using decorative finishes additionally to the plaster, wallboard or other surface, *please* check the flame-resistant properties of the materials used. Some plastic materials ignite easily and/or give off fumes and such materials should *never* be used for decorative wall finishes.

INTERNAL FINISHES – FLOORS

In previous chapters we have talked about screeds, floating floors, wood block and wood strip flooring. These finishes are just as suitable in commercial building work as they are in domestic construction.

In case study 1.2, however, we mentioned power floating of the structural concrete. This technique, originating from America, has proved successful as a means of eliminating the need for screeded floor finishes, which are primarily of a corrective nature. Neither finish, in itself, produces the final working surface and we can now look at some floor finishes that do provide this.

Example 3.8

A library floor has to provide a sound-deadening surface that can be easily maintained. What is the most suitable finished surface?

It would be possible to use a cellular backed vinyl sheet floor covering with cold welded seams to provide a continuous surface that can be easily cleaned. This surface would, however, be subject to indentations and abrasion from street shoes, and could deteriorate quite quickly.

An alternative is broadloom carpet but there would be wastage in cutting around book shelving and reference filing cabinets. Wear would also be uneven with pedestrian traffic using a primary route across the floor more fully than in the reading areas.

A finish that provides the same qualities as the carpet and underlay is woollen carpet tiles on a bonded ribbed rubber base. The ribbing prevents sliding and also improves sound deadening, and the carpet reduces both airborne and impact noise. The tile format (450 or 500 mm^2) allows for minimum cutting wastage and also provides for some areas of the floor to be interchanged with others to balance the wear. The surface can be vacuumed without problems of lifting or displacement and offers a warm visual appearance.

Example 3.9

A supermarket floor has to provide a durable surface that can be cleaned easily and will not stain from spillage of goods such as red wine, blackcurrant jam or salad cream. It must also be capable of sustaining display stands of a temporary nature without suffering indentations. What finish will be most suitable?

In order for the floor to be cleaned easily and resist stains, the surface must be smooth and impervious to liquid. The logical choice, therefore, is some form of clay quarry tiling of sufficient thickness to resist cracking from articles such as bottles being dropped on its surface. The tiles should be bedded in a sand—lime—cement mortar and pointed with a weaker mix of the same materials. They should have a polished surface rather than a glazed one to reduce the risk of crazing (fine cracking) from normal pedestrian traffic and trolleys.

Example 3.10

A gymnasium/sports-hall floor has to accommodate a variety of activities including badminton, basketball and general gymnastics. What is the most logical floor finish?

A gymnasium floor needs to be sprung slightly to avoid a 'dead' feeling from its usage. It would be wrong, therefore, to use parquet or wood block flooring even though these are sometimes seen in halls and general usage school buildings. The only way that some flexure can be provided is by using tongued and grooved hardwood strip flooring supported by timber battens. The battens in turn are clipped into place, the clips having a rubber sandwich under the battens. The hardwood selected should be close grained and of high density. Philippine mahogany, for instance, would not be suitable, being coarse grained and of low density, which would result in rapid wear and deterioration.

EXTERNAL FINISHES – ROOFS

This topic has been covered fairly extensively in Volume 1 and to some degree in chapter ·1 of this book. The variations of roof finish for flat roofs, in particular, can be developed a little further to illustrate how the problems inherent in flat roof construction can be overcome.

Flat roofs, particularly those of domestic properties, need to develop a high thermal insulation property. Currently the required U value is about 60% of that specified for the walls.

The roof must resist the passage of moisture into the building, and the conventional materials used to coat the roof are built-up bituminous felt and asphalt. As discussed before, both these materials are susceptible to temperature changes and should be protected. Because the roof is flat it may be subject to a fair amount of access. It may be that the surface needs to accommodate pedestrian traffic. Rain bounces off a roof to the same degree as it does off a pathway, and protection is required at parapet walls. Also the rain-water needs to be drained away in most instances, so that gutters should be designed for durability and ease of maintenance.

A traditional flat roof has its insulation located either in the roof space (if using timber joists) or on the underside (if using concrete).

If the insulation is transferred to the upper side, it can be located either below or above the felt or asphalt.

Sandwich construction

Here the insulation (for example, rigid expanded polystyrene sheets) is located under the felt or asphalt. In order to produce a surface compatible with the laying of the felt/asphalt, a light mesh and screed should be laid over the insulation. The conventional treatment of the felt/asphalt deck is to use white spar chippings to provide solar reflection. Two other alternatives, which give protection to the felt/asphalt and provide a more consistent temperature, are:

(a) to provide a constant depth of water (e.g. 50 mm or more): this keeps the membrane cool and also protects it from drifting in strong winds;

(b) to provide topsoil, to a depth of 50 mm, which is then turfed. In this treatment it is wise to provide an ash bed below the topsoil to allow roof drainage and to deter the growth of moss.

[*Note*: Whichever technique is used, it will be necessary to provide a vapour barrier below the polystyrene.]

Inverted insulation construction

If the insulation is placed *above* the felt/asphalt this acts as both the membrane and the vapour barrier. We have already discussed the use of bituminised lightweight aggregate to form this insulation layer. If rigid expanded polystyrene sheets are used instead, they will need to be weighted down to prevent them from blowing away. Concrete paving slabs may be used or stone chippings (65 mm thickness), or a combination of the two, with the slabs (50 mm thick) acting as a check, or curb, to the chippings.

Figure 3.4 shows the various treatments described here, with details of the treatment at the parapet wall.

EXTERNAL FINISHES – WALLS

These have also been dealt with in Volume 1 and in chapter 1 of this book. In this chapter we shall content ourselves with the self finishes that can be achieved for reinforced concrete.

The fact that we are considering concrete is, in itself, something of an anomaly since it is notoriously poor as an insulating material. It is, however, a popular material with designers by virtue of the wide variety of finishes that can be produced. In examining these

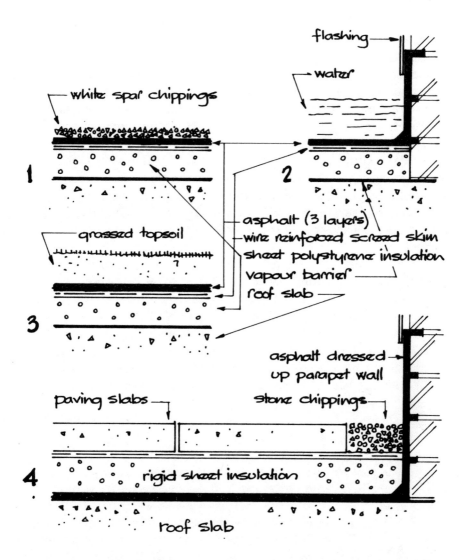

flashing

water

white spar chippings

grassed topsoil

1

2

3

asphalt (3 layers)
wire reinforced screed skim
sheet polystyrene insulation
vapour barrier
roof slab

asphalt dressed up parapet wall

paving slabs

stone chippings

4

rigid sheet insulation

roof slab

Fig 3.4 DETAILS OF FINISHES TO ASPHALTED CONCRETE ROOF SLAB SHOWING SANDWICH (1 TO 3) AND INVERTED (4) INSULATION FORMATS

finishes, we need to divide the material into two categories, precast panels and in situ walling.

Precast panels

These may take the form of cladding panels between columns or permanent shuttering to walls, and in both cases are cast in the horizontal plane. The finish can therefore be created in this horizontal plane with much better control. Also the surface is not exposed to severe atmospheric conditions and can be steam cured. Where the surface is heavily textured, it should be realised that the reinforcement cover must be to the thinnest section. For thin slabs such cover requirement can have a serious effect on the thickness, making it up to two-and-a-half times the required structural thickness. Such cover requirements can be reduced dramatically if stainless steel rather than mild steel reinforcement is utilised, since the former will not corrode.

A texture that has proved popular by virtue of its weathering characteristics is random vertical striations, these being produced by combing the concrete in its plastic state. When located as cladding these fine ridges assist the rain-water to run off the surface and clean the recessed areas of the concrete, thus heightening the effect. If, however, the striations run horizontally, they form irregular ledges to trap the rain-water and stain the surface. This should be avoided as the building becomes unsightly within a fairly short space of time. Smooth, patterned surfaces, using geometric shapes, should also be avoided since the planes that tilt upwards gather grime and those that tilt sideways become blotchy. A fine, smooth surface is also likely to develop hair cracks in which dirt can be trapped.

Another texture that utilises specially selected aggregate for the surface only is 'exposed aggregate'. The large aggregate (a popular choice is granite) is laid in the shutter tray on a water-based adhesive to retain it in position. The reinforcement is then located and the concrete placed and vibrated to remove air voids. As the concrete hydrates it drives off its water content, which breaks up the adhesive so that the selected aggregate loses its bond with the shutter but retains its key to the concrete panel. After the panel is struck it is hosed down with a high-pressure spray, which accentuates the aggregate in relief. An alternative process, using the normal aggregate of the concrete, is to spray the shutter with a retarder. In this way, after placing the reinforcement and concrete, and vibrating it and striking the shutters, the surface can be hosed and wire brushed to remove

the still-plastic surface of cement and fine aggregate. For details of
these finishes see fig. 3.5.

In situ concrete
Although it is possible to use retarders in the shutters for in situ
work, they tend to run down the surface producing a greater depth
of exposure at the base than at the top of the shutter. It is much
more satisfactory to produce 'shuttered' finishes that do not actually
affect the surface of the concrete but produce a variety of relief
effects.

A very dramatic effect, suitable for very large areas such as under-
pass walls, is obtained by producing a series of glass-fibre or plastic
moulds in a numbered sequence. The perimeter joints are accentuated
but the panels form a pattern that relates to adjacent panels to
produce a relief mural. This can be repeated on the opposite wall

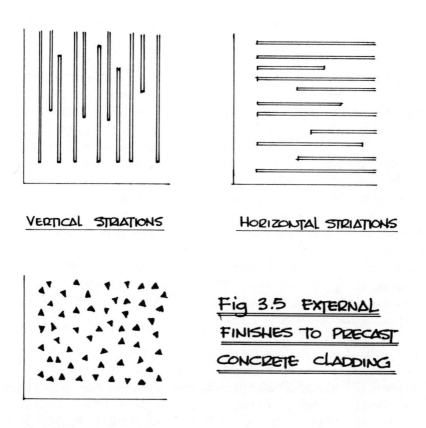

VERTICAL STRIATIONS   HORIZONTAL STRIATIONS

EXPOSED AGGREGATE

Fig 3.5 EXTERNAL
FINISHES TO PRECAST
CONCRETE CLADDING

and gives the impression that the surface is a series of preformed panels located carefully into their positions.

A second finish that gives visual depth to the surface is a 'random board' treatment. Instead of using a ply shutter, a series of lengths of timber of differing widths and thicknesses are fixed to heavy cross-battens to form the shutter face. The shutter is sprayed with a generous layer of mould oil so that it will release easily from the concrete when struck. The texture is then echoed in the concrete surface. This texture is most effective in close-up and should be reserved for ground level or elevated walkway areas. It is wasted, for example, on lift motor rooms or third-floor elevations.

A surface that relies even more heavily on close-up inspection is 'sawn board' finish. Instead of using differing thickness of board, a consistent thickness is battened together. The other variation is to use rough sawn timber so that the mould oil accentuates the grain and natural defects such as knots, which are then translated to the concrete surface. As with the striated panels, both sawn and random board finishes are best suited to vertical alignment if they are to weather satisfactorily.

A very expensive treatment, because of its high labour content, but one that is rather impressive, is bush-hammered concrete. A technique that gives a good effect is the use of corrugated iron shuttering; after its removal the raised surface is bush hammered to expose the broken aggregate while retaining the smooth surface in the recessed zone. An excellent example of this can be seen at the elephant house in London's Regent's Park Zoo.

The examples described here are illustrated in fig 3.6. There are many more treatments for both in situ and precast concrete, too numerous to detail in this one chapter. A wealth of background reading including colour photographs is readily available.

Rather than consider horizontal surfaces (ground slabs) in this chapter, these are best dealt with in chapter 6 in association with external works, since this will include an element of 'soft' and 'hard' landscaping.

Assignment 3.1

A sports pavilion comprising dressing-rooms, shower area, bar, kitchen and main hall is built in cavity wall construction, with latticed steel beams supporting a flat roof. Select the internal finishes for each zone, giving a reasoned explanation as to their suitability and with a

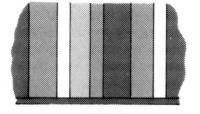

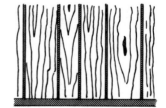

random board shutter

sawn board shutter

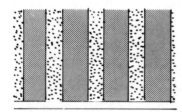

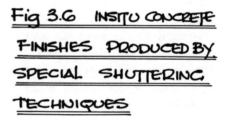

Fig 3.6 INSITU CONCRETE FINISHES PRODUCED BY SPECIAL SHUTTERING TECHNIQUES

corrugated iron shutter
(bush hammered ribs)

view to initial cost and cost-in-use characteristics. Use of trade journals is recommended in answering this assignment.

Assignment 3.2
A two-storey school building comprising a gymnasium with an art studio above is constructed using steel frames with a low-pitched portal roof. Select a suitable external treatment to the building and sketch details of the wall and roof finishes.

4 Internal Fittings

Fittings might be described loosely as those parts of a building that serve a necessary function but do not contribute to its structure or general fabric. The obvious domestic areas are sanitary appliances, kitchen units, bedroom units, and storage shelving. Stairs might also be considered in this category since they fit the general description quite well.

It should be noted that in all cases these fittings may be manufactured in factories by specialist suppliers, in either finished or component-assembly form. Because of this, it is very rare for a builder to produce such components since the costs would usually exceed those of the mass-produced items. Work on site would be restricted to the assembly and fitting, hence their name. Therefore, although it is probably good for the soul to know how the various fittings are manufactured, it does not form part of the technician's role. What is important is an understanding of the performance, materials, required work space (or circulation space), associated services (as applicable) and fitting. This should be coupled with an appreciation of standard sizes, modular co-ordination and physically restricting dimensions.

STAIRS

Stair flights can be produced in a variety of layouts but have certain common physical restricting dimensions. Let us consider the requirements for a domestic stairway.

The width of the stair, the component going and rise dimensions and handrail height are set out in the Building Regulations. While you need to ensure that the minimum and maximum dimensions are not exceeded it is unlikely that a standard stair flight, supplied by a joinery manufacturer, would fail to conform. By way of illustration, a standard straight flight comprises thirteen rises of 200 mm each (to tie in with a storey height of 2600 mm) with goings at 225 mm. This

produces a pitch of 41.6°. The width, inside strings, is 800 mm and
the handrailing has an effective height of 850 mm. Thus, all Building
Regulations limits are met.

The limiting dimensions that cannot be pre-set by standard joinery
are those of headroom clearance and length of landings, which are
dependent on the design layout. These factors are shown in fig 4.1,
which also illustrates the minimum allowances that have to be made
for movement of furniture, such as a double wardrobe. Where stairs
change direction, the width of half- and quarter-landings, should be
at least that of the stairs.

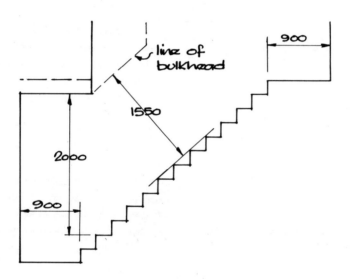

Fig 4.1   MINIMUM DIMENSIONS RELATING
TO HEADROOM AND CIRCULATION AROUND STAIRS

The layout is very much the province of the designer and should
make the best use of the space available. Straight flights are not
always desirable and it is sometimes convenient to 'fold' a stair in
order to release more space for bedrooms or similar.

Examples of typical plan layouts are shown in fig. 4.2.

Stair flights other than the conventional softwood tread-and-riser
type are available but, with the possible exception of the hardwood
open-rise straight flight, are extremely expensive by comparison.
Spiral, open string and central spine staircases often use mild steel

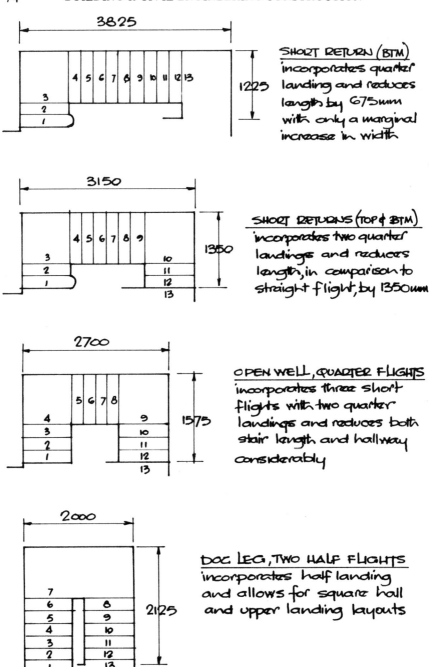

SHORT RETURN (BTM) incorporates quarter landing and reduces lengths by 675mm with only a marginal increase in width

SHORT RETURNS (TOP & BTM) incorporates two quarter landings and reduces length, in comparison to straight flight, by 1350mm

OPEN WELL, QUARTER FLIGHTS incorporates three short flights with two quarter landings and reduces both stair length and hallway considerably

DOG LEG, TWO HALF FLIGHTS incorporates half landing and allows for square hall and upper landing layouts

Fig 4.2 ALTERNATIVE STAIRCASE LAYOUTS

(sometimes chromed) in conjunction with hardwood treads. It is worth remembering, however, that most people like to carpet their stairs and this can be achieved most easily on conventional 'tread-and-riser' flights, whatever their layout.

SANITARY APPLIANCES

Domestic fittings are generally restricted to six items: bath, wash-hand basin, water-closet, bidet, shower tray (or cubicle), and sink. These are available in a variety of materials and often in a variety of sizes and shapes. Because of this it would be sensible to consider each item in turn and to examine their associated space requirements.

Bath

The four most commonly available materials are cast iron, pressed steel, acrylic and glass fibre. The cast iron bath benefits from its rigidity, in that it can stand free, and its ability to retain heat. It is more expensive than other materials but is very popular with users. Pressed steel offers a reasonable alternative but is of lighter gauge and requires some bracing. Acrylic baths do not have the rigidity of cast iron and require timber sub-frames for bracing. However, they do have some advantages over cast iron and pressed metal. They are not coated with enamel, the colour being consistent for the total thickness; they may be produced in a wide range of mouldings, including non-slip patterned base; and they are light to handle and install. It is also easier to match shades of side and panels, since these are usually acrylic regardless of the bath material. Glass-fibre baths are slightly cheaper than acrylic but do not have the smoothness of finish or the colour consistency. Because of this they are not very popular with users.

Sizes available for all types include 1700 x 700 mm, 1700 x 760 mm, 1800 x 800 mm, 1070 x 690 mm and 1500 x 1500 mm. The last two sizes are respectively for 'sit-in' baths (primarily for old people) and corner baths with a diagonal working area. The 1700 x 700 mm module is by far the most readily available and because of this presents the best value for money. Also side and end panels are available in a wide variety of pressings. The associated space require-ment (for drying) is 700 x 1050 mm and this may affect the size and layout of the bathroom.

Wash-hand basin

By far the most commonly available material is vitreous china. However, some basins are available in acrylic, for use in built-in vanitory units. Most basins are available in pedestal and wall-mounted format, using the same unit but with different fixings. Alternatives are corner basins (for use in cloakrooms), countertop (for partial building in) and bowl shaped (for fitting into vanitory units).

Sizes vary considerably but as a general rule those for pedestal/wall fixing are larger (average size 635 x 510 mm), those for wall fixing alone slightly smaller (average size 510 x 405 mm), corner basins (400 x 400 mm), countertop similar to pedestal (610 x 510 mm) and bowl types similar to wall fixing only (550 x 400 mm). These are average figures and different manufacturers may vary either dimension in increments of 5 mm. The reason for lack of modular sizing is that very few basins are built in and where they are it only requires a small amount of tile cutting. The space requirement, however, is consistent, being 1100 mm width x 650 mm back from the basin, allowing sufficient space for face washing. [*Note*: For handwashing only, in cloakrooms, the width may be reduced to 750 mm.]

Water-closet

The W.C. basin (or pan) is available only in vitreous china but the cistern, if coupled to form a suite, may be in vitreous china or acrylic. Alternatively, a concealed cistern may be made of heavy-duty polythene of similar quality to that used for cold-water storage tanks.

The sizes will vary depending on the format of basin and cistern, average dimensions being 730 x 520 mm for a 'close coupled' suite, 650 x 550 mm for a basin coupled to a slim cistern (acrylic) and 500 x 280 mm for the basin alone. The overall heights vary (800 mm for close coupled and 940 mm for open coupled) but the pan height is constant at 405 mm. To allow flexibility of siting, the discharge may be specified to either side or directly to the rear of, for ground floor use, direct downward outlet.

Space requirements forward of the pan (550 mm) and width (750 mm) should take account of other fittings. Since overlaps of functional space are permissible when fittings are alongside each other, the overall width of a bathroom will not necessarily be the sum of the individual fitting requirements but will exceed their physical dimensions (see fig. 4.3 for layouts).

Bidet

As with the W.C. basin, the only readily available material is vitreous china. Dimensions are fairly consistent at 540 x 350 x 380 mm

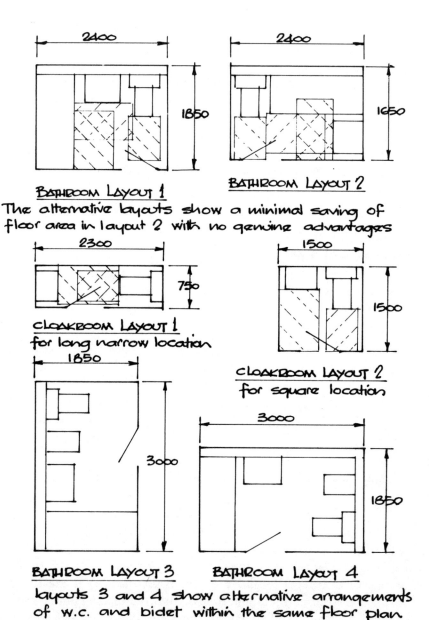

2400

1850

BATHROOM LAYOUT 1

2400

1650

BATHROOM LAYOUT 2

The alternative layouts show a minimal saving of floor area in layout 2 with no genuine advantages

2300

750

CLOAKROOM LAYOUT 1
for long narrow location

1500

1500

CLOAKROOM LAYOUT 2
for square location

1850

3000

BATHROOM LAYOUT 3

3000

1850

BATHROOM LAYOUT 4

layouts 3 and 4 show alternative arrangements of w.c. and bidet within the same floor plan

Fig 4.3 TYPICAL BATHROOM & CLOAKROOM LAYOUTS SHOWING ALTERNATIVE SITINGS OF APPLIANCES – SHADED AREAS SHOW FUNCTIONAL SPACE REQUIREMENTS

height and space requirements are 550 x 700 mm. This fitting is available in two basic forms, over-rim water supply (this uses taps in the same way as wash-hand basins and can utilise the common hot and cold water supplies); and through-rim water supply, possibly supplemented by a spray (as the name implies the supply is inside the fitting and requires separate supplies for hot and cold water to avoid the risk of contaminating the supplies to other fittings. See fig. 4.4 for diagrammatic line drawing of installation.

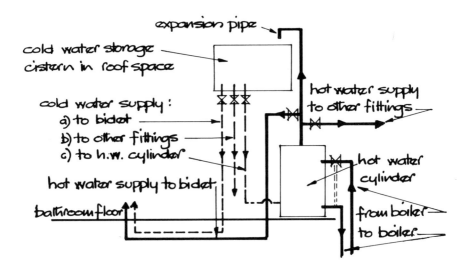

Fig 4.4 DIAGRAM OF SEPARATE HOT AND COLD WATER SUPPLY TO "RIM FEED" BIDET

Shower tray
Glazed fire-clay is the traditional material and also the most stable form of shower tray. Alternative materials such as acrylic plastic or pressed metal need considerable bracing to avoid racking under load. Even with this there will inevitably be some distortion, which can affect the perimeter seals and allow shower water to track behind the tray.

Three sizes are available, namely 600, 750 and 900 mm². In practice the smallest of these would be awkward to use, since 750 mm is the recommended space for washing. An alternative to the shower tray, if acrylic is preferred, is the shower cubicle, which is cast in one piece to include the shower spray and controls. With the shower tray a cubicle or recess has to be constructed and tiled and provided with a door or curtain.

Sink

The traditional sink uses the same material as the shower tray, namely fire-clay, but this type of sink is rarely specified in housing nowadays except possibly for a laundry/utility room. Modern kitchens are fitted with sink units, which match the remainder of the kitchen, having a sink top, which either lips over the base unit or is set into the worktop.

Materials available are cast iron — enamelled; pressed steel — enamelled; stainless steel; acrylic; and glass fibre. At one time the pressed steel enamelled sink was popular but as stainless steel sinks became more competitive in price the white enamelled sink lost favour. However, the inset bowl, using coloured enamels, is making a recovery in more expensive kitchens. Acrylic and glass-fibre sinks are not very durable in comparison when considering the aggressive effect of cutlery. The most popular units by far are those made from stainless steel. Sizes are related to the kitchen units, the built-in sinks being smaller than the 'sink tops', and are all based on a depth of 600 mm with widths of 1000, 1200 and 1500 mm.

Circulation space is better dealt with in the overall consideration of fitted kitchens. At this point we shall concentrate on the common factor between the six items of sanitary fittings we have just examined. How do the fittings connect to (a) the supply and (b) the wastes?

In chapter 2 we dealt with the hot and cold water supplies and the internal sanitation and drainage. We saw there the general layout of each so we need only look at the detail. This can best be shown diagrammatically. (See fig. 4.5 for supplies and fig. 4.6 for wastes.)

KITCHEN UNITS

This area of house construction more than any other has captured the interest of interior designers and, by a long-running advertising campaign, has been 'sold' to the householder. Whereas people might consider some minor modification to the structure of their house or the building of an extension after long deliberation, they will quite happily spend thousands of pounds on refitting a kitchen even if it functions perfectly well.

Leaving aside the glossy magazines, kitchen showrooms and the offer of 'total design' made by kitchen 'specialists', the main factor

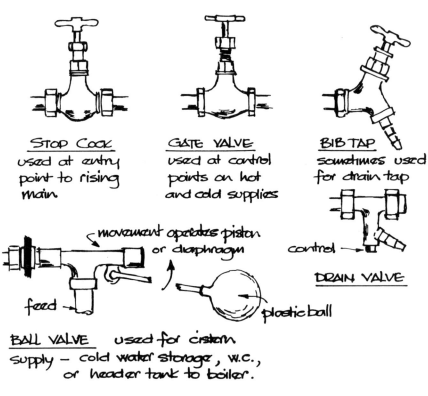

STOP COCK
used at entry
point to rising
main.

GATE VALVE
used at control
points on hot
and cold supplies

BIB TAP
sometimes used
for drain tap

movement operates piston
or diaphragm

control —

DRAIN VALVE

feed —

plastic ball

BALL VALVE   used for cistern
supply — cold water storage, w.c.,
or header tank to boiler.

SINGLE TAP
used for sink, bath
or wash hand basin

MIXER TAPS
used as for single
taps inc. use with bidet

Fig 4.5   FORMS OF CONTROL COMMON

IN   HOT AND COLD WATER SUPPLY SYSTEMS

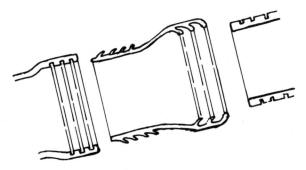

W.C. CONNECTOR

shown in cross section, this plastic connector grips
over the outlet of the w.c. pan and into the drainage
system. It can be used for all materials of pipework

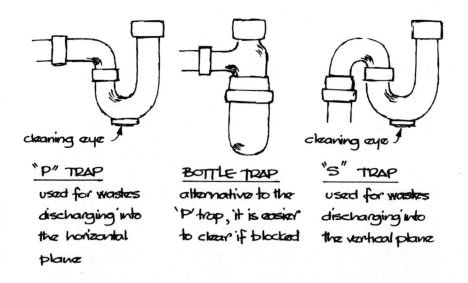

cleaning eye

cleaning eye

"P" TRAP
used for wastes
discharging into
the horizontal
plane

BOTTLE TRAP
alternative to the
'P' trap, it is easier
to clear if blocked

"S" TRAP
used for wastes
discharging into
the vertical plane

Fig 4.6   TYPICAL DOMESTIC WASTE CONNECTIONS
(ALL TRAPS PROVIDE WATER SEAL, W.C. TRAP BUILT-IN)

is the development in mass-production techniques, with the wide variety of surface textures, colours and details that can make a kitchen unique. It would be wrong to attempt to cover, in one section of a chapter, the full range of details, since these can be obtained easily from a product information book supplied by any individual manufacturer.

What we can examine is the concept of the fitted kitchen and apply it to some worked examples. Units fall into four basic categories:

(a) *Base units*, including provision for sinks, hobs, and tables (depth 600 mm including worktop).
(b) *Midway units* to hold spice jars or locate power points (depth 100 mm).
(c) *Wall units*, including upper units, cooker hoods and spot lighting (depth 300 mm).
(d) *Tall units*, including food cupboards, broom storage, provision for building-in of cookers, fridges and microwave ovens (depth 600 mm).

They all come in a variety of lengths, based on a 100 mm module, in multiples of 300, 400, 500 and 600 mm, i.e. 300 and 600; 400 and 800; 500, 1000 and 1500; 600 and 1200; except the tall units, which are generally restricted to the 500 and 600 mm sizes but also include a 700 mm special for built-in appliances only. Figure 4.7 shows the interrelationships of the modules. The modules can be tied visually by a 'running' worktop, which will span several base units. Some manufacturers produce these, to order, for changes in direction with no joints. This worktop link can be quite useful if a pair of controlling walls do not quite fit the modules (common enough in existing properties), where the gap between units can be filled at its base to form a tray recess.

Example 4.1
Provide a suitable kitchen layout for the house shown in fig. 1.12 in Volume 1. This will be two straight runs of 3.6 m and will include two tall units (one to house the oven and one to house the refrigerator).

The front wall includes a window. This wall should therefore house:

(a) the sink unit − because drainage is easier on an external wall;
(b) the hob unit − because the window will aid ventilation and because it is near the sink unit.

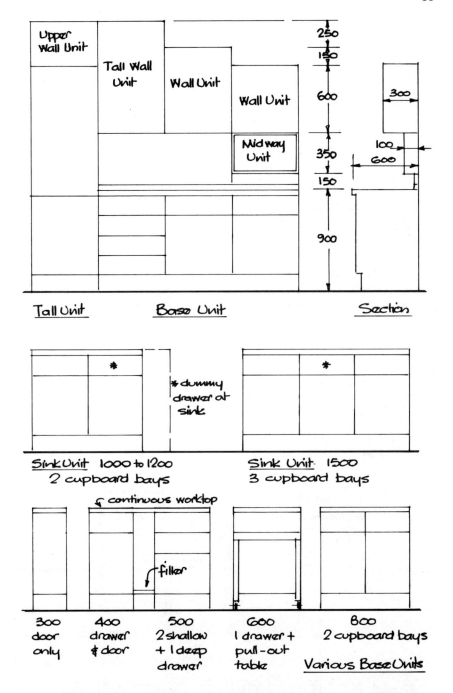

Fig 4.7 TYPICAL KITCHEN UNIT MODULES

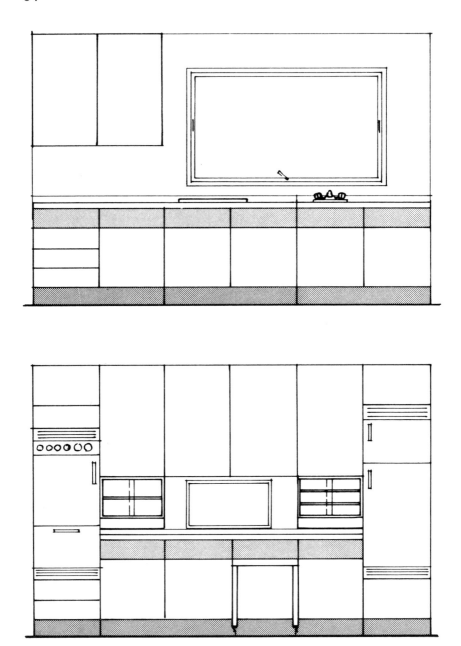

Fig 4.8 FITTED KITCHEN TO EXAMPLE 4.1

The rear wall has a service hatch to the dining room and the space between the two banks of kitchen units is 1.5 m. This wall should therefore house:

(a)   the tall units at opposite ends to keep oven and fridge apart (the latter will in fact be a fridge/freezer);
(b)   a pull-out table to the base units;
(c)   wall units − to run between the tall units;
(d)   midway units on each flank of the service hatch to the dining room − these can house the spices since the table can also act as a food preparation area linking food, sink and cooker.

The results are shown in fig. 4.8.

Example 4.2
Provide a suitable kitchen layout for the ground-floor flat shown in fig. 5.8, Volume 1. The cooker is to be free standing (electric with hood over) and the refrigerator is to fit below the worktop area.

The sink is located on the front wall, as explained in example 4.1. The flank wall can contain the cooker with wall units to include the cooker hood. A pull-out table can also be provided in this run of units with the stools housed behind it. The fridge is to be located on the rear wall. The overall dimensions are 3750 x 2100 mm so there will need to be some infilling. Also the corners are continuous so there will need to be special corner base units. The results are shown in fig. 4.9.

BEDROOM UNITS

Until fairly recently, bedroom units as such were rare. Some bedrooms had small built-in wardrobes utilising standard doors with a built-up door lining plus a shelf with a hanging rail below. Others had no provision at all since house owners, by tradition, bought wardrobes and dressing tables. However, the success of the kitchen-unit manufacturers 'rubbed off' into the bedroom furniture market. Joinery manufacturers saw that mass-production techniques and modular co-ordination could be applied equally well to bedroom furniture.

The initial development was to use similar casing components to those of kitchen 'tall units' for the sides and rear and to fit louvred doors to the front. This allowed for shelved areas as well as hanging

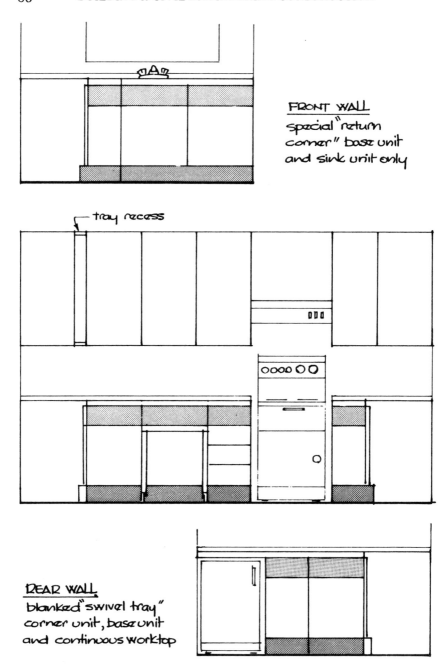

FRONT WALL
special "return
corner" base unit
and sink unit only

tray recess

REAR WALL
blanked "swivel tray"
corner unit, base unit
and continuous worktop

Fig 4.9 FITTED KITCHEN TO EXAMPLE 4.2

rails. From this beginning the full range has developed to include dressing tables, vanitory units with wash-basins, drawer units and tall areas split into drawers, shelves and hanging space. However, whereas the kitchen units have now become part of the new house package, regardless of price range, fitted bedrooms are still often to low specification, the bulk of the high specification work being commissioned by the householder some time after moving in. An example of what might be achieved is shown in fig. 4.10. A more usual form is shown in fig. 4.11.

STORAGE SHELVING

This is very much the province of the commercial building sector rather than domestic building. It can be divided into two areas, free standing, made up from slotted steel angles (which is *not* a fitting); and built-in shelving for storage and/or display (which *is* a fitting). In this chapter, therefore, we will consider only the latter.

Display areas include tilted shelves, drawers, horizontal shelves, glass cabinets and base cabinets. The last of these is virtually identical to some of the base units described in kitchen work. The remainder are generally based on the system of slotted channel and locking bracket supports. These are formed of mild steel, which is enamel painted in black, white or grey. Black goes well with wood veneered fittings while white matches quite well with melamine-faced particleboard work. Grey is usually restricted to workrooms and storage areas. Figure 4.12 shows the system as applied to a menswear shop:

All of the fittings discussed require some services back-up. Sanitary appliances require water supply and drainage pipework. Kitchen units require both of these and additionally gas and electricity. Bedroom fittings require electricity supplies and sometimes water and drainage. Shop fittings require electricity supplies for spotlighting. It is important to make this provision at the preliminary stage prior to installing the fittings, since provision of supplies afterwards could lead to problems and could result in damage to the units. As with other construction phases, plan ahead in a logical manner to solve problems before they occur. This makes for a happier work-force, less frustration and a superior end result at an economic price.

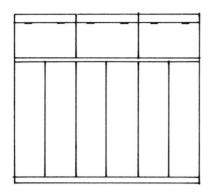

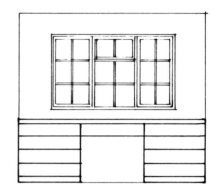

Combination of three
hanging wardrobes with
three top-hung cupboards

Combination of Two base
unit drawer sets with a
linking worktop dressing
table unit

MAIN BEDROOM — Units on opposite flank walls

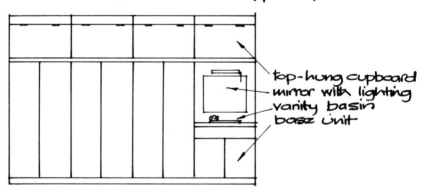

top-hung cupboard
mirror with lighting
vanity basin
base unit

Combination of three hanging
wardrobes with top-hung cupboards
plus vanitory base unit with
top-hung cupboard lining through

SECOND BEDROOM
Units on long wall

Fig 4.10 EXAMPLE OF FULL PROVISION OF
FITTED BEDROOM UNITS FOR MASTER BEDROOM
(WITH EN SUITE BATHROOM) AND SECONDARY
BEDROOMS (TO INCLUDE WASHING FACILITIES)

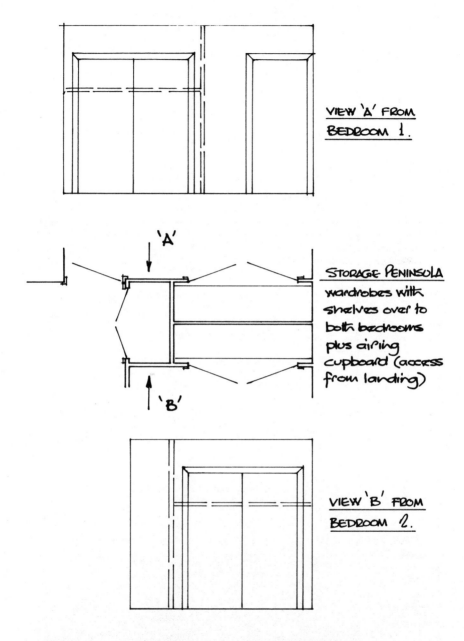

VIEW 'A' FROM
BEDROOM 1.

'A'

STORAGE PENINSULA
wardrobes with
shelves over to
both bedrooms
plus airing
cupboard (access
from landing)

'B'

VIEW 'B' FROM
BEDROOM 2.

Fig 4.11   EXAMPLE OF TYPICAL STORAGE
FITTINGS TO TWO MAIN BEDROOMS

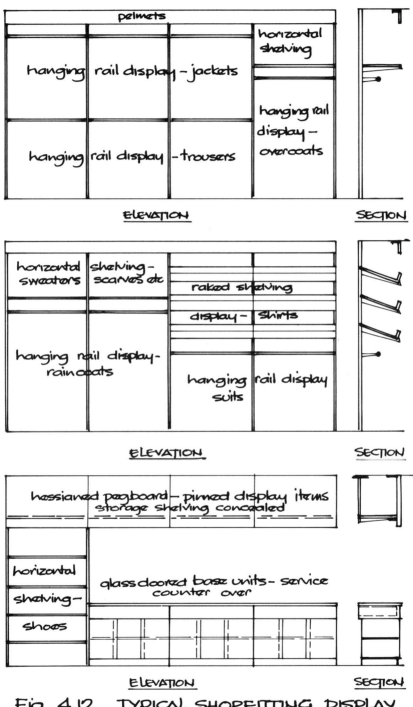

pelmets

hanging rail display – jackets

horizontal shelving

hanging rail display – overcoats

hanging rail display – trousers

ELEVATION

SECTION

horizontal shelving – sweaters

shelving – scarves etc

raked shelving

display – shirts

hanging rail display – raincoats

hanging rail display suits

ELEVATION

SECTION

hessianed pegboard – pinned display items
storage shelving concealed

horizontal shelving – shoes

glass doored base units – service counter over

ELEVATION

SECTION

Fig 4.12   TYPICAL SHOPFITTING DISPLAY
SYSTEM BASED ON 1200 mm MODULE

Assignment 4.1
Produce a section through the stair well (fig. 5.16, Volume 1) to show half-landing, baluster and headroom dimensions, assuming a storey height of 2.6 m.

Assignment 4.2
Select suitable sanitary ware for the en-suite bathroom to the master bedroom (fig. 5.16, Volume 1) and show the arrangement of waste runs to the vertical drainage stack. Use a manufacturer's catalogue and produce a rough costing of the associated taps, wastes and plugs.

Assignment 4.3
Produce a fitted kitchen layout (fig. 5.16, Volume 1) to a high specification. Draw up a plan and elevations and write a brief specification of the equipment. It will be necessary to use a manufacturer's catalogue for this assignment.

5 Groundworks

The term 'groundworks' is at first sight a little confusing, since it involves work that is primarily below ground rather than at ground level. In practice the construction work below and up to ground level is related either to buildings, in which case it is called 'the substructure'; or to large-scale works like roads or railways, in which case it is called 'external works'. Groundworks are associated with both of these and concern the state of the ground to be worked from the point of view of: investigation; ground-water; slope stability; foundation principles, with recommendations for types of retaining wall or foundation; and the consideration of settlement. At this level of study we shall confine ourselves to the work associated with low-rise construction and small-scale external works.

SITE INVESTIGATION – PRELIMINARY

A large number of people in the construction industry think of site investigation purely as boreholes and laboratory tests. These are part of the exercise, of course, but they do not dominate the investigation. At the preliminary stage a lot of information can be gained by researching the area that is to be developed. After all, the ground has been there a long time and it is unlikely that it has never been utlised in its entire history.

We can start by constructing a history of the site going back to the times of the industrial revolution. This will help to establish the purpose to which the site might have been put and can be determined by visiting the archives section of the local library. Examples of the type of activity that might have taken place are:

(a) *Mining*: Has the ground ever been prospected for tin, coal, other minerals?

(b) *Tunnelling*: Have exploratory or practice tunnels ever been excavated?

(c) *Previous building*: Have other buildings been constructed, and subsequently demolished, with some basement work or heavy foundations left in position?

(d) *Other uses*: Has the site ever been used as a dumping ground and subsequently backfilled and levelled?

(e) *Forestation*: Has the site been recently cleared of trees or bushes?

All of these could affect the cost and method of construction for the proposed development. (a) and (b) would pose the risk of sudden ground collapse! (c) would add expense in breaking out the previous work plus the subsequent consolidation of backfill. (d) could pose problems of both chemical action and settlement on loose backfill, and (e) could affect the water table, causing ground heave.

Following this, the drainage division of the local authority should be contacted to establish:

(a) whether there are any drainage works — sewers, surface-water drains, agricultural drains crossing the site;

(b) whether there are any storm drains, underground streams or springs, artesian wells on the site.

The presence of (a) could affect the drainage system for the proposed works, whereas (b) could mean huge costs to divert natural water courses in order to make the development possible!

The next stage is to enquire of the building control or district surveyor's office as to problems that may have been encountered in the recent construction of any adjacent property

(a) What was the ground-water level?

(b) What were the subsoil characteristics?

(c) What sort of bearing pressures were used?

This will consolidate the background picture of the site further in determining whether, in fact, it is worth developing or if the extra work will make the substructure costs prohibitive.

Finally, the scope, nature, magnitude, and location of any work on or adjacent to the site should be determined from the various service authorities. This might affect the location of the buildings and will probably be of help for connections for the proposed work.

SITE INVESTIGATION – PHYSICAL

This may be carried out either by the main contractor or by a specialist subcontractor, to confirm the suspicions developed from the preliminary investigation. It is likely that a specialist investigation will only be needed for medium- and high-rise work, since the type of construction considered in low-rise is not generally likely to produce exceptionally heavy loadings. Because the walls or foundations are likely to be fairly simple, the work is generally carried out by the main contractor.

Let us look at what information we need to have. It is necessary to know what type of soil and subsoils we are likely to meet or load with the structure, so we need to go to about three times the depth of the foundations. This can be done by using a *hand auger*, which is drilled into the ground manually. Cores of soil samples can be taken at intervals and inspected for their composition and density. This should also give us an idea of the ground-water level when inspecting the samples and by 'sounding' the borehole.

It would be useful to know the likely angle of internal friction of the subsoils to the maximum depth of excavation. This will help us to establish the types of trenching support we will need to adopt. A series of *trial pits* will show this and confirm the level of ground-water if it is at a level above proposed foundation work. If the pits are left open for a while, we can also see the effect of exposure on the subsoil, which could pose problems; this is particularly true of chalky soils.

An estimate of the permissible bearing capacity of the subsoil is also needed to establish foundation sizes. Two site tests will help to establish this. *Plate loading* at the base of the trial pit will give a fair indication of this but will necessitate medium-term loading of a test rig. The plate should also be of a reasonable size, which means that the load will have to be fairly large. It is best used as a confirmation that the design stress is not likely to be exceeded. However, it has an added benefit in that it also measures short-term settlement.

The other technique is to use penetration tests to measure the rate at which the subsoil can be penetrated and to translate this into bearing capacity. The *standard penetration test* involves driving a tube, using a drop hammer to a standard force, into the ground and measuring the number of blows against the rate of penetration. Alternatives are the *dynamic cone test*, where the tube is replaced by a cone, and the *Dutch cone test*, where the cone is rammed hydraulically into the ground rather than hammered. If the required loading rates are very low, e.g. on a road rather than a building, a variation of the plate loading test using a smaller plate can be performed. This is the *California bearing ratio test*, which can utilise a vehicle as the reaction to the applied load. It does, however, give an empirical (rule-of-thumb) result rather than an exact figure on which to work.

Finally, it is necessary to know of any peculiar chemical properties that are present. This has to be established under laboratory conditions and so some of the sample taken from the hand auger should be packaged into plastic containers and sent to a laboratory for analysis at an early stage. This may affect the type of cement to be used, particularly if there are sulphates in the soil.

For more extensive construction work the use of a specialist site investigation firm is recommended. This will involve the drilling of boreholes and the analysis of the samples of unconfined, confined and undisturbed soils under laboratory conditions. The more common tests include shearbox and triaxial compression tests but there are many others. The results are recorded and built into a 'site investigation report'. This aspect of groundworks is considered more fully at higher technician level under the discipline of soil mechanics.

GROUND-WATER CONTROL

Having talked about the possible problems of ground-water, what is it and how can we control it? If all subsoils were granular, either sand or gravel, water would drain away. This is why there are deserts. Fortunately, this is not generally the case and in the U.K. a wide mixture of subsoils guarantees that this will not happen. Layers of clay prevent surface-water from draining away completely and if the clay layer forms a bowl, it can trap the water, even in sandy/gravel soils to some depth. Water from large rivers also permeates the subsoils to quite a distance from the natural river bed. These two sources − rain-water and river-water − combine to form ground-water.

How we dispose of it will depend, to a large extent, on how much of it there is to control. A small amount in one excavation may be controlled by digging a *sump* (a deeper pit at one point) and installing a pump, driven by a compressor, to discharge the water into the surface-water drainage system. Alternatively, if working close to a river, it may be necessary to use *steel sheet piling* to contain the site, again adopting the principle of pumping out the area. The majority of work, however, lies somewhere between these two extremes and is concerned with fairly large areas of site where the ground-water is only a problem during excavation. In these cases it is possible to *dewater* the contained area by a circuit of *well points*. Each well point comprises a perforated cylinder with a 'jetting' head. Inside the cylinder is a tube with a fine meshed screen and a non-return ball valve. The well points are connected to hoses feeding a ring circuit hose connected to a compressor.

Prior to connecting up the circuit, each point is jetted down through the granular soil under high water pressure passing round the tube and through the head. When the complete circuit of well points is in position and connected up, the compressor pump is put into reverse and ground-water is drawn off continuously thus lowering the water table inside the circuit. On completion of the affected work the compressor can be disconnected and the water table will re-establish itself. On large sites with varying levels, separate circuits at correspondingly different depths can be formed with each discharging into either a surface-water drainage system or a nearby stream or canal.

The methods described here are all concerned with the temporary removal of ground-water. If a permanent solution is required, this can be achieved by other methods but this will change the character of the subsoil and is removal rather than control. This is outside the immediate scope of construction at this level and is more appropriate to higher award studies. Figures 5.1 to 5.3 show the ground-water control techniques described in preceding paragraphs.

SLOPE STABILITY

Slope stability was introduced in chapter 7 of Volume 1 when we looked at the effect of ground-water on the internal friction of soil particles. As explained previously, the exclusion of such ground-water will affect the character of the soil so we must look at other methods of achieving the stability of slopes.

The most obvious technique is to assess the most critical angle of internal friction and make the slopes, at changes of level, fall to a more gentle angle of repose. By installing land drains a short distance behind the slope, with convenient discharge ducts, we can reduce the risks of land slip still further. However, in freak rainstorm conditions the slope might still fail but the likelihood of total failure is very small.

Some sites are very exposed and this can lead to wind and rain eroding the slopes at changes of level if the soil is unprotected. It is sensible, then, to protect the surface against such erosion but how can this best be achieved?

A simple but ugly solution is to spray the slope with concrete,

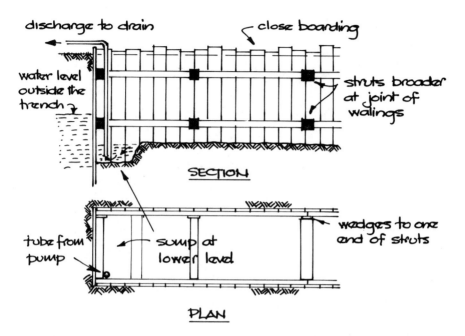

discharge to drain

close boarding

water level outside the trench

struts broader at joint of walings

SECTION

tube from pump

sump at lower level

wedges to one end of struts

PLAN

Fig 5.1 PUMPED SUMP METHOD OF GROUND WATER CONTROL IN TRENCH

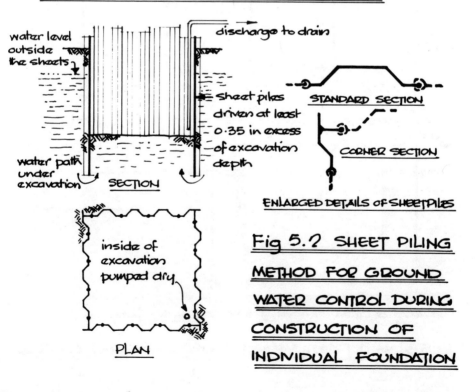

water level outside the sheets

discharge to drain

sheet piles driven at least 0.35 in excess of excavation depth

STANDARD SECTION

CORNER SECTION

water path under excavation

SECTION

ENLARGED DETAILS OF SHEETPILES

inside of excavation pumped dry

PLAN

Fig 5.2 SHEET PILING METHOD FOR GROUND WATER CONTROL DURING CONSTRUCTION OF INDIVIDUAL FOUNDATION

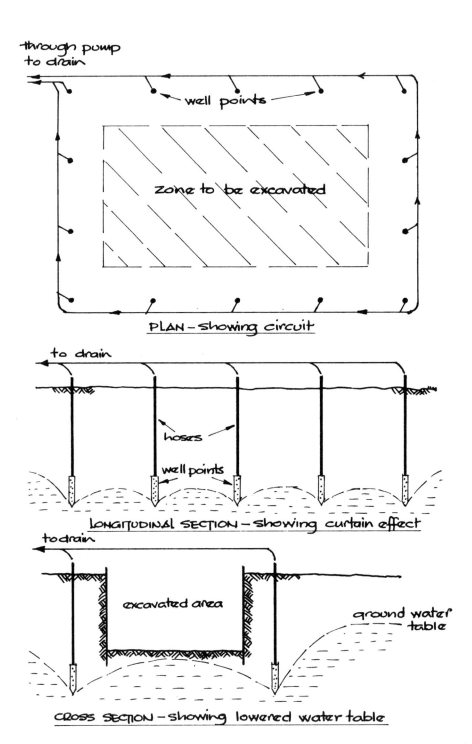

through pump
to drain

well points

zone to be excavated

PLAN — showing circuit

to drain

hoses

well points

LONGITUDINAL SECTION — showing curtain effect

to drain

excavated area

ground water
table

CROSS SECTION — showing lowered water table

Fig 5.3 WELL POINT METHOD OF GROUNDWATER CONTROL

using small aggregate and cement mixed dry and pumped under pressure through a hose, the nozzle of which feeds a metered water supply to the mix.

A slightly more satisfactory technique is to carpet the slope with precast concrete units of a high relief texture or to form an interlocking pattern of partly open concrete units so that some soil can still be grassed.

The most preferred 'environmental' solution, if functionally acceptable, is to plant shrubs, which will bind the soil and establish a network of roots to help contain the ground-water.

Figures 5.4 and 5.5 illustrate the second and third of these techniques.

RETAINING WALLS

From a practical consideration, the most durable way to stabilise a change of level is by using retaining walls. We examined mass-retaining walls in Volume 1, fig. 7.3, and saw how the middle third law dictated the shape and mass of material necessary for stability. Where long runs of wall occur this may prove very expensive on both materials and labour. We shall now look at a more economic use of materials employing reinforced concrete retaining walls. These can be used to resist considerable ground pressures. However, we shall concern ourselves with walls retaining earth to a depth of only 3 m or less.

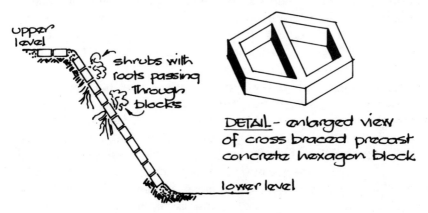

upper level

shrubs with roots passing through blocks

DETAIL - enlarged view of cross braced precast concrete hexagon block

lower level

Fig 5.4 EMBANKMENT STRENGTHENED WITH KEYED HEXAGON BLOCKS PROVIDING FOR SHRUBS AND GRASSED FINISH

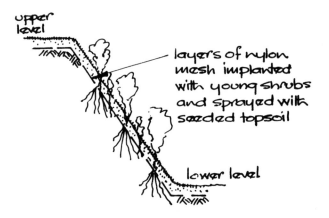

upper level

layers of nylon mesh implanted with young shrubs and sprayed with seeded topsoil

lower level

Fig 5.5 EMBANKMENT STRENGTHENED WITH NYLON MESH SHEETING CONTAINING SHRUBS TO PROVIDE STABILISING INFLUENCE

A major benefit of reinforced concrete is that the tension face can be predetermined by the location of the reinforcement. We can, therefore, make the wall either a buttress or a tension structure (see fig. 5.6). The buttress type of wall is quite useful for small retaining structures, since it demands a minimum of excavation. The tension structure as an alternative is more efficient, since it uses the mass of the earth that it is retaining; because this is backfilled after the wall has been built, it permits the inclusion of land drains and weepholes for walls that do not need to be watertight.

A freestanding wall of the types discussed here may be either a *cantilever* or a *counterforted* structure, the former term meaning that all the force can be taken by the stem of the wall, and the latter that it can be distributed through stiffening ribs located at intervals along its length. Counterforting is not very economic at heights of up to 1.5 m but becomes more economic than the plain cantilever as the height of retained earth is increased. Figure 5.7 shows typical examples of both forms of wall, acting as either buttress or tension structures.

Assuming that the reinforced concrete has the correct properties to resist failure by cracking, what are the other ways in which a reinforced concrete retaining wall might fail?

The force being retained might be too great to be catered for by

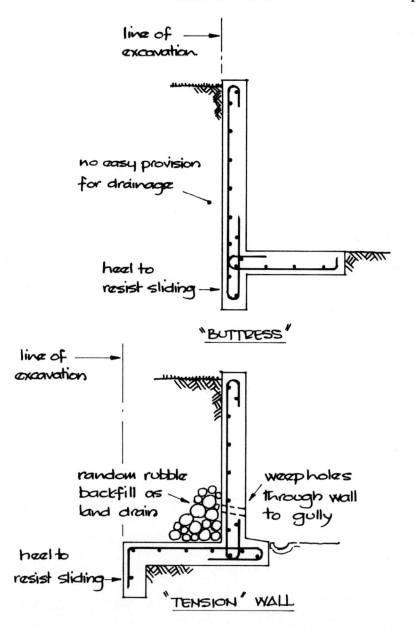

line of excavation

no easy provision for drainage

heel to resist sliding

"BUTTRESS"

line of excavation

random rubble backfill as land drain

weep holes through wall to gully

heel to resist sliding

"TENSION" WALL

Fig 5.6 SIMPLE REINFORCED CONCRETE CANTILEVER RETAINING WALLS – IDEAL FOR WALLS AROUND 1.5 TO 2.0m HEIGHT

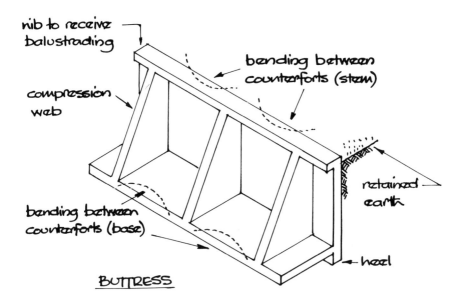

BUTTRESS

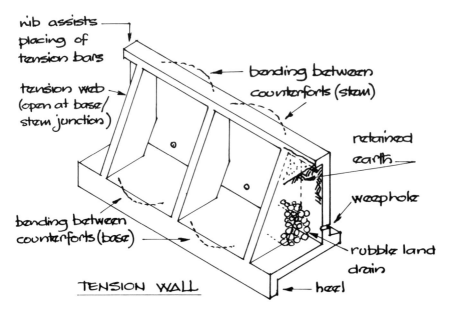

TENSION WALL

Fig 5.7 COUNTERFORTED R.C. RETAINING
WALLS - SUITABLE FOR WALLS ABOVE 2·0m
HEIGHT

frictional resistance and the wall might slide. This is best dealt with by constructing the base of the wall a sufficient depth below the lower ground level so that the soil at the lower level provides 'passive resistance'. Alternatively, a downstand nib may be provided to form a 'key' with the subsoil.

A second failure mode is overturning. The active moment of the retained earth acting at its centre of pressure may be greater than the gravity of the wall and any contributory mass or force acting through its centre of gravity. This is less likely to occur on a tension structure than it is on a buttress.

Because the nature of earth is variable it is normal to apply a 'safety factor', whereby the passive resistance to sliding and the passive resistance to overturning exceed their active counterparts in the ratio 1.5:1. These effects are shown in fig. 5.8. (Further discussion on the behaviour of retaining walls will be found in *Construction Science*, Volume 2 by Granada).

Not all retaining walls are built just to provide changes of level. Some form an integral part of the substructure to a building, e.g. a basement. Good use can be made of the building itself in providing additional strength to such retaining walls. In this case the wall need not act as a cantilever or counterfort structure since the ground-floor slab, at the top of the retaining wall, can provide a 'propping' effect as a direct resistance to the horizontal force of the retained earth. This type of wall is known as a *propped cantilever* and is illustrated in fig. 5.9. [*Note*: Special care needs to be taken with this type of wall to ensure that the ground slab has gained sufficient strength before backfilling the earth to the face of the wall.]

FOUNDATION PRINCIPLES

In the early part of the chapter we talked about site investigation and the need to determine soil properties. This is so that foundations of an economic size can be provided to transmit the loads from the structure safely through the subsoils without the building, bridge or tower (for example) sinking into the ground. We also looked briefly at this aspect in Volume 1 and in the examples worked here we will refer to table 2.2 of that book.

We know that in mass concrete foundations the load from a wall or column is transmitted through a plane of 45° to the underside of the base. Any concrete that lies outside this line will develop tension because of the resistance of the earth to the imposed load and will

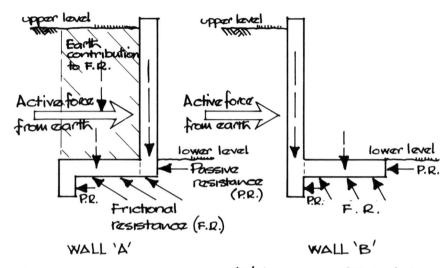

WALL 'A'   WALL 'B'

<u>SLIDING RESISTANCE</u> – Wall 'A' is more efficient than
Wall 'B' since it utilises the gravitational force of the
earth above the wall, in addition to that of the wall
itself, to produce the frictional resistance component

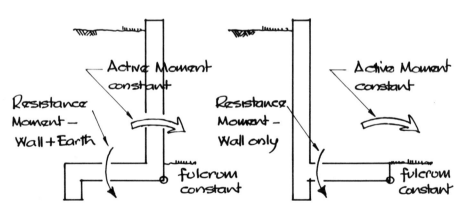

<u>RESISTANCE TO OVERTURNING</u> – Wall 'A' more efficient

<u>Fig 5.8   PASSIVE RESISTANCE TO SLIDING AND
OVERTURNING FOR TYPICAL CANTILEVER WALLS</u>

Exaggerated deflected forms for cantilever
and propped cantilever retaining walls

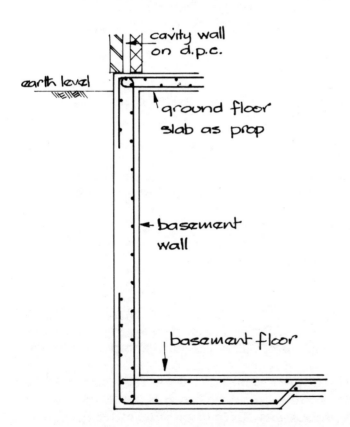

Fig 5.9   STRUCTURAL BEHAVIOUR AND
CONSTRUCTION OF PROPPED CANTILEVER

fail. Thus the plan area of the base must lie inside this 'shear plane'. This means that, if the base size is determined by bearing capacity, the 45° spread will dictate its thickness. If the thickness of concrete is impractical or uneconomic it will need to be reinforced to resist the tension, which, because of the upward earth pressure, will be on the bottom face of the foundation. This is explained diagrammatically in fig. 5.10 for both mass and reinforced single-column foundations.

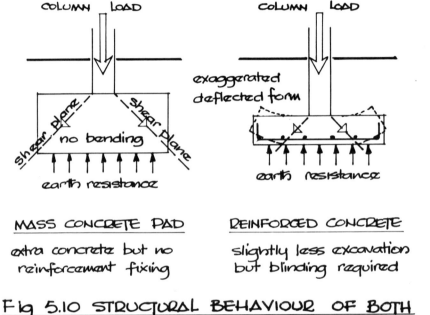

MASS CONCRETE PAD

extra concrete but no
reinforcement fixing

REINFORCED CONCRETE

slightly less excavation
but blinding required

Fig 5.10 STRUCTURAL BEHAVIOUR OF BOTH MASS AND REINFORCED CONCRETE FOUNDATIONS

Example 5.1
A three-storey office building of framed construction is built on a subsoil of firm sandy clay. Design a foundation for a typical internal column having dimensions of 300 x 300 mm and carrying a load of 500 kN.

Assume a bearing capacity of $100-200$ kN/m^2, confirmed by plate loading test as 180 kN/m^2. The load does not take account of the ground slab, the soil overburden or the mass of the foundation itself (see fig. 5.11). Assuming this as 25 kN/m^2 (generous), we therefore modify the permissible bearing capacity to 155 kN/m^2.

Since load/area = stress we know that by transposing we can find the area required, i.e. load/stress = area. Thus $500 \div 155 = 3.226$m^2,

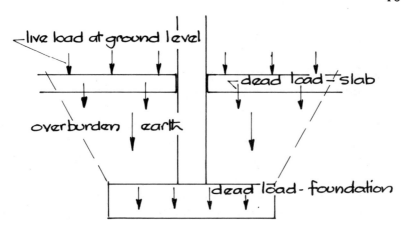

Fig 5.11 LOADS, ADDITIONAL TO THOSE FROM THE COLUMN, TO WHICH THE GROUND UNDER THE FOUNDATION IS SUBJECTED

which for a square foundation gives a base size of $\sqrt{3.226} = 1.796$ m. Use 1.8 x 1.8 m base. Taking the 45° spread line from the face of the column, this will produce a base thickness of $0.9 - 0.15 = 0.75$ m or 750 mm.

This is not particularly impractical and might well be economically acceptable. Suppose, however, that this had not been the case. How could we have determined the thickness for a reinforced concrete base?

In practice a reinforced concrete foundation should never fail in tension, provided that it has sufficient tensile reinforcement. Its depth will rather be dictated by shear, with the column trying to push its way through the base and the ground pressure pushing around the shear perimeter of the column. Suddenly to introduce the use of Code of Practice 110, with talk about ultimate loads and characteristic strengths, would be confusing in this book and the detail will be covered in *Construction Science* Volume 2. However, we can say here that the result will be a reduction in the overall thickness.

[*Note*: Concrete for structural foundations is stronger than that for normal domestic strip footings and mixes of 1:2:4 and 1:1½:3 are not uncommon.]

The example that we examined was for an isolated column. This is quite normal when a building is constructed on a square grid. However, not all structures follow such a format. Where columns are located in fairly close proximity in one plane of the grid it might well be sensible to use *combined footings*, i.e. to provide one common foundation for two columns. Figure 5.12 illustrates this for two columns with equal loading. If all columns carried equal loads, combined foundation design would produce no problems. Unfortunately, this is not the usual case and one column will normally be carrying more load than its immediate neighbour.

In cases where column loads differ for a combined footing, the base will not be symmetrical about both columns but will take one of three forms:

(a) *Trapezoidal*, beloved by many writers of construction books but not very sensible in practice, from either the point of view of design or that of fabrication. This is because the calculation for centre of gravity is complex; shuttering for sides of foundation is awkward; and all lateral reinforcement is of different lengths.

(b) *Rectangular*, preferred by designers and contractors as being simpler to produce. Calculation is not required in order to find centre of gravity for a rectangle; shuttering is straightforward; and reinforcing is consistent.

(c) *Cantilever*, useful on occasion but more complex than the rectangular type. Here, beam design includes calculation of shear stirrups to support the cantilevered column; shuttering involves base and beam with provision of a compressible layer under the cantilever; reinforcement includes shear stirrups and a more complex arrangement of bars.

All three of these combined footings are known as *balanced foundations* since the centre of gravity of the foundation is balanced to coincide with the centre of force of the column loads.

Since the rectangular format is the most popular type let us examine its application in an example.

Example 5.2

A three-storey furniture warehouse has a rectangular column grid of 6 x 3 m. Design a combined footing to carry column loads of 550 kN and 420 kN on the 300 mm² reinforced concrete columns. The subsoil is of sandy clay and has a confirmed bearing capacity of 160 kN/m².

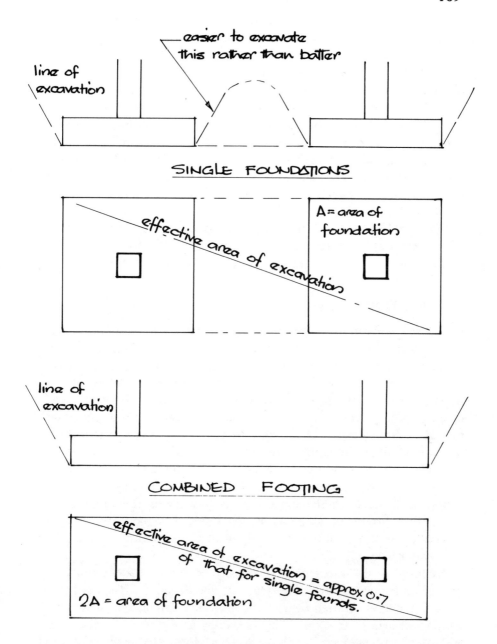

easier to excavate
this rather than batter

line of
excavation

SINGLE FOUNDATIONS

effective area of excavation

A = area of
foundation

line of
excavation

COMBINED FOOTING

effective area of excavation = approx 0.7
of that for single founds.

2A = area of foundation

Fig 5.12 COMBINED FOOTING SHOWING SAVINGS
IN EXCAVATION & SHUTTERING OVER SINGLE PADS

Allow for an overburden (to include self weight of the base) of 30 kN/m². This will reduce the effective bearing capacity to 130 kN/m², giving a required base area of $(550 + 420) \div 130 = 7.46$ m². Use 7.5 m² as design area. Determine centre of force for columns by taking moments about one of them: $550 \times 3 = 970 \times \overline{x}$. Therefore $\overline{x} = 1650 \div 970 = 1.7$ m from the 420 kN load (or 1.3 m from the 550 kN load). Locate the centre of the base at this point, i.e. use a 5 x 1.5 m base (7.5 m²) 2.5 m to each side of the centre of force, to extend 0.8 m past the centre of the 420 kN load and 1.2 m past the centre of the 550 kN load, and 0.75 m to each side of the column grid. Suggest reinforced concrete since the thickness for mass concrete would be 1.05 m.

Figure 5.13 shows the three types of balanced foundation and fig. 5.14 illustrates the base designed in example 5.2, showing how the reinforcement would be arranged for a reinforced concrete base.

SETTLEMENT PROBLEMS

In low-rise construction, settlement problems are not commonplace, being more likely to occur for heavily loaded buildings. However, settlement does occur and can, if not catered for, create problems.

As discussed at the beginning of the chapter, the physical site investigation should provide adequate information about the soil characteristics. Clay soil, as we discussed in Volume 1, contains pores of water trapped between the clay particles. If the clay is subjected to heavy loading, these pores are compressed and the water is squeezed out. This process produces settlement under the load, which can be split into two categories, initial settlement and long-term settlement. Initial settlement, as its name implies, occurs as the building or structure is produced. In contrast, long-term settlement occurs over a number of years of the functioning of the structure until *all* the pore water is expelled.

As long as the settlement is consistent for all foundations, no problems associated with the structure should occur. If some foundations settle more than others, the structure can become distorted, which will lead to cracking. This is known as *differential settlement* and should be avoided at all costs. One factor that will help is to design all foundations as close to a common bearing pressure as possible. Thus it is important not to round up too generously on odd foundations.

A second factor, which is one of precaution, is to tie the isolated

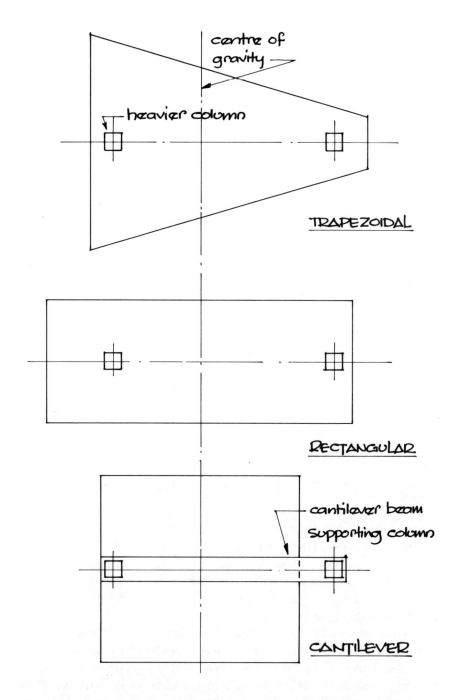

centre of gravity

heavier column

TRAPEZOIDAL

RECTANGULAR

cantilever beam supporting column

CANTILEVER

Fig 5.13 THREE ALTERNATIVE FORMS OF
BALANCED FOUNDATIONS FOR UNEQUAL COLUMNS

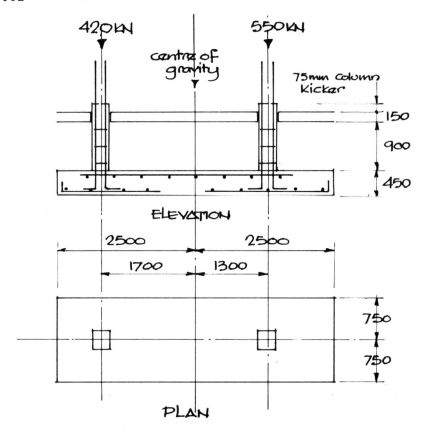

Fig 5.14   RECTANGULAR BALANCED FOUNDATION CONSIDERED IN EXAMPLE 5.2

foundations together with ground beams. This will stiffen the foundation zone against racking and reduce the risk of differential movement. [*Note*: This is primarily a problem in clay or silty clay soils. It should not occur in granular soils such as sand so no precautionary ground beams would be needed.]

In some instances it is essential to minimise the problem of settlement even when there is no likelihood of differential movement. In such cases the answer is to 'pre-load' the foundation, simulating loading in excess of that to be carried, and allowing periods of partial recovery during the pre-load. The physical mass if using ballast such as kentledge would be huge and it is more practical to employ ground anchors as a resistance against which to jack the pre-load.

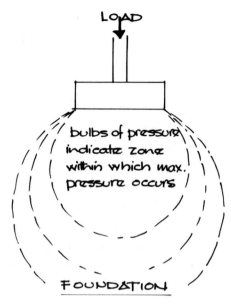

LOAD

bulbs of pressure
indicate zone
within which max.
pressure occurs

FOUNDATION

pores of water in clay
– no pressure

compressed pores under
load – water squeezed out

SOIL BEHAVIOUR

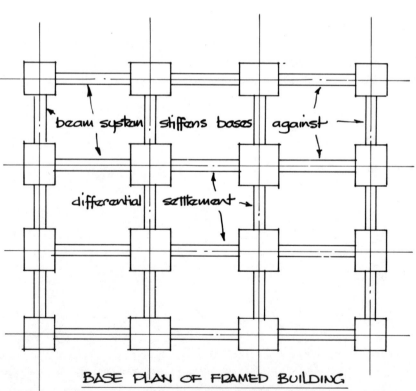

beam system stiffens bases against

differential settlement

BASE PLAN OF FRAMED BUILDING

Fig 5.15 PRINCIPLES OF FOUNDATION SETTLEMENT
AND PRACTICE TO MINIMISE DIFFERENTIAL MOVEMENT

Figure 5.15 shows the principles and practice involved in settlement of foundations.

Assignment 5.1
Using a piece of land local to you as the basis of a site to be developed, carry out a preliminary site investigation based on the principles outlined at the beginning of this chapter. [*Note*: You should explain the purpose of your enquiries when talking to the local and service authorities.]

Assignment 5.2
Assume a site has dimensions of 50 x 20 m and has a water table that needs to be reduced by 0.7 m while foundations are being constructed. Using data from suppliers' specifications for well points and pump handling capabilities, design a layout for a ring circuit to dewater the site. Discharge is to a nearby stream.

Assignment 5.3
Design a mass concrete foundation to carry a column of a three-storey department store. The column size is 400 x 400 m and the load is 1100 kN. The subsoil is of dry sandy/gravel mixture.

6 External Works

The term 'external works' is an all-embracing description of the entire range of operations that does not involve the construction of buildings. Major works such as roads, railways, associated bridges and open culverts for canal work are the more noticeable aspect of this work. Culverting under roads, embankment protection and drainage services are less obvious, and earthwork and its various problems are probably the least well appreciated aspect of external works. This is because if earthwork is well executed nobody notices it.

EARTHWORKS

It is very rare when dealing with large sites, such as those for road and rail work, to find the ideal conditions of subsoil and land contours. We talked, in the last chapter, about control of ground-water for small sites, but think of the problem it can pose on sites that spread over several hectares.

It is quite common to find that a fault in the substrate, where a clay layer has sheared, will cause a fairly widespread area of trapped ground-water. This might cause problems for the proposed work and also during the construction period if not removed. A well-established technique used to correct the problem is the sinking of sand drains through the clay substrate into the granular soil layers below it. A mechanical auger (mobile drilling rig) excavates boreholes, which are then filled with sand. The ground-water drains through this sand and the trapped water table is removed. The size and spacing of these sand drains will depend on the nature of soil below the clay and the volume of water to be controlled. This is a permanent solution to the problem, and is necessary because water will feed into the area via the fault in the clay layer at a fairly consistent rate. (See fig. 6.1).

Fortunately, such a large volume of ground-water is not common to all large sites. For those where no permanent solution is required it may be sufficient to use caterpillar-tracked equipment and ignore

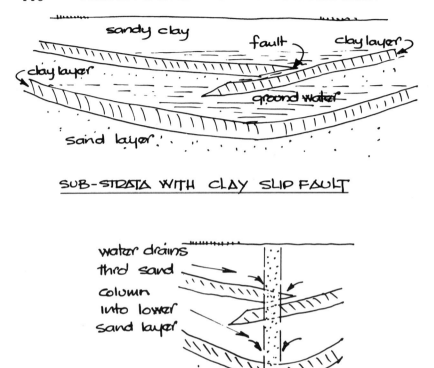

SUB-STRATA WITH CLAY SLIP FAULT

Fig 6.1 TECHNIQUE FOR REMOVING TRAPPED GROUNDWATER BY SINKING SAND DRAINS

the mud. The type of equipment used is often somewhat larger than that associated with small construction sites. A few of the composite face shovel/backhoe type of excavator are, however, used, but are 'tracked' rather than 'wheeled'. The larger equipment is normally of single function and the 'dozer' is fairly common.

In Volume 1 we talked about 'cut and fill' and considered its application in simple terms. For large open sites we need to extend our thinking. Although the principles still hold good, the execution is of a much larger scale and needs to be calculated as part of the survey using 'mass haul' techniques. This is considered in higher award studies. It is sufficient to know at this stage that wherever possible the 'spoil' and 'fill' should be approximately equal. If not, the contractor will either need to 'cart away' huge volumes of spoil (expenses on haulage and dumping) or 'import' huge volumes of fill (expenses on purchase and haulage).

When we considered groundworks in chapter 5, we discussed the preliminary site investigation. When this is related to large-scale sites the added preliminary search of geological maps should be undertaken. This should indicate the type of soils likely to be encountered: rock outcrops, which might need to be 'broken out'; clay beds indicating seasonal movement; and, more seriously, large deposits of limestone or chalk. This last factor could mean that a system of underground springs, wells and lakes exists in potholes and swallow-holes, since running water cuts easily through these soft rocks. As far as pothole systems are concerned, they will probably have been mapped quite reliably, exploration of shafts, tunnels and caves being a popular pastime. Swallowholes are a different case and may be located by sonic survey techniques.

Ideally the location of these underground hazards will have been established at a sufficiently early stage to allow for planned re-routing of the road or railway. Occasionally, however, isolated swallowholes may need to be dealt with, this being more expedient than relocating the road or railway. The only certain way of curing the problem in this instance is to fill the swallowhole with an inert material that is self-consolidating. Pulverised fuel ash (P.F.A.) is the ideal material since it can be pumped to its location. You might ask, 'Why re-route the road when it is possibly more logical to re-route an underground stream?' The answer is that although it may be possible to culvert a stream this should only be considered either (a) if the stream is very small, or (b) if there is no other practical solution, since the water has been coursing through the subsoil for centuries and probably has smaller 'overload' channels through the rock, which may feed in or out only in extreme conditions. These may not all be located in a survey and will inevitably cause problems to the completed work, possibly years after construction has finished.

EMBANKMENTS AND CUTTINGS

The layout of roads and railways, particularly the latter, should produce a reasonably level route of 'carriageway' or 'permanent way'. Gradients of rail systems are often in the order 1:250 or less, to avoid undue stress being put on the engine or brakes of the large number of trains travelling the route. To achieve this, the construction will often involve the formation of cuttings and embankments. These can be quite extensive and will be considered on a large scale in higher award studies, both in the construction and soil mechanics areas of work. Let us then consider the 'medium-scale' treatment of this external work to a depth or height of around 3 m.

A *cutting* is formed by removing the earth local to the road or railway, by cutting it away to produce sloping sides at the natural slope of the ground, or almost vertical sides using retaining walls. The cutting can be formed using a hydraulically operated backhoe, since this type of machine can operate quite efficiently to a height or depth of 4.0 m. The backhoe excavator should work in conjunction with spoil lorries, which will transfer the excavated soil to the site of the embankment. Depending on the length of cutting and the slope of the ground, the excavation can be done directly along the route, digging from below, or from above the proposed route. [*Note*: Cuttings should not be formed for long distances since drainage of the surface-water could pose handling problems.] The sub-grade of the road can be formed using bulldozers and angle-bladed graders but on large-scale works scraper lorries would be more efficient. Ground drains will need to be laid to both sides of the carriageway to take the surface-water from the slopes as well as that from the road.

The stability of the slopes, as discussed in chapter 5, involves the siting of land drains and discharge gullies. These gullies can take the form of surface-mounted rock drains. This involves cutting channels into the slope and backfilling with large, broken aggregate, including flint and other durable rock types. These are run at an angle to the slope rather than vertically, to slow down the surface-water to speeds that can be handled by the drainage runs. (See fig. 6.2.)

An *embankment* is formed by building up the area local to the proposed road to sustain the loads likely to be imposed on it. Depending on the type of soil excavated from the cuttings, the embankment construction might be by one of three commonly applied methods:

(a)   fill graded and consolidated in layers, possibly using bitumen binders to stabilise the soil;
(b)   forming a central area of palleted wire cages (gabions) of rock bound with weak cement/sand grout: this is then banked up at each side with the spoil;

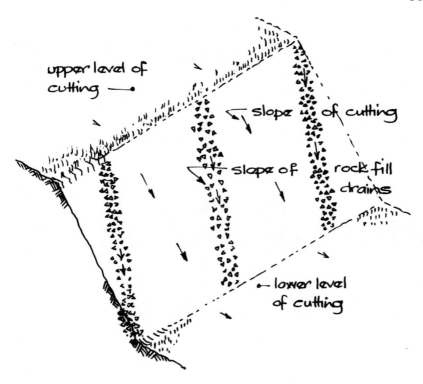

upper level of
cutting ——•

slope of cutting

slope of

rock fill
drains

lower level
of cutting

Fig 6.2 SURFACE DRAINAGE OF CUTTINGS

BY ROCK FILLED CHANNELS

(c) constructing trestle support frames to an elevated section of
road and then banking up to it with the spoil.

These methods minimise the risk of settlement under operational
conditions. Where consolidation is carried out, this should be done
first using a 'sheep's foot' roller, followed by a vibrating smooth-
surfaced roller. Slope stability treatment should follow the format
described in chapter 5 for both cuttings and embankments but the
drainage is only essential to the cuttings as far as road drains are
concerned.

Consolidation of fill may be necessary for some parts of the road
where weak pockets of earth occur, and this can take the same form
as that described for the base of the road to the embankments. It
might also be useful to stabilise the area with bitumen injected into
the subsoil. (Figure 6.3 and 6.4 show details.)

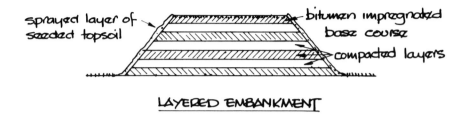

sprayed layer of seeded topsoil

bitumen impregnated base course

compacted layers

LAYERED EMBANKMENT

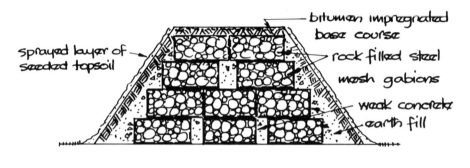

sprayed layer of seeded topsoil

bitumen impregnated base course

rock filled steel mesh gabions

weak concrete

earth fill

GABION EMBANKMENT

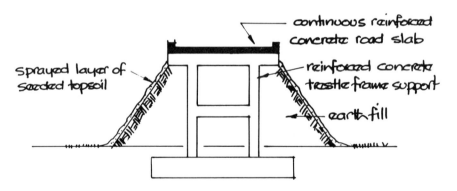

continuous reinforced concrete road slab

sprayed layer of seeded topsoil

reinforced concrete trestle frame support

earth fill

TRESTLED EMBANKMENT

Fig 6.3 ALTERNATIVE METHODS OF FORMING EMBANKMENTS FOR RAIL OR ROAD WORKS

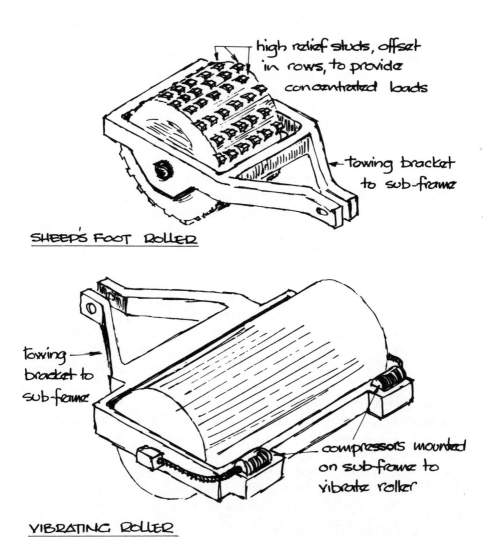

high relief studs, offset in rows, to provide concentrated loads

towing bracket to sub-frame

SHEEP'S FOOT ROLLER

towing bracket to sub-frame

compressors mounted on sub-frame to vibrate roller

VIBRATING ROLLER

Fig 6.4 HEAVY MECHANICAL EQUIPMENT USED FOR CONSOLIDATION OF RAIL/ROAD SUB-GRADES

CULVERTS AND HEADINGS

Occasionally it may be necessary to cross below the road, for either service runs or pedestrian safety. If such a crossing point occurs at an embankment, the culvert should be built prior to the construction of the embankment using reinforced concrete cast in situ in formwork. If, however, the crossing point is below the general ground level, ramps should be formed to the reduced level prior to its formation. It might still be logical to construct the crossing point using 'cut and cover' techniques if the road has not yet been commissioned, but for crossing points below an existing road this would not be practical. The cut and cover form of culvert involves sinking sheet piles, or trench sheeting for less deep excavation, to form temporary sides to the crossing point. As the earth is excavated the open area is propped and the earth levelled to the required depth. [*Note*: Sheet piles can act as cantilever retaining walls if driven to a sufficient depth. The additional length of penetration required is 0.35 times the required depth.]

The base slab and wall kickers are cast and the wall shutters located and braced. The walls are then cast, after which the roof slab is constructed. When the concrete strength is sufficient, the shuttering and trench supports are removed and the sides of the culvert back-filled. The ground can then be reinstated. (See fig. 6.5 for construction details.)

The alternative to culverting, when crossing below existing roads (or railways) is 'headings'. The crossing comprises sections in reinforced precast concrete with their edges forming mechanical jointing keys. The sections are jacked through the soil (using a cutting edge on the leading unit) with excavation cleared back through the heading. Obviously the jacks need a resistance against which to work. This is provided to one side of the embankment by casting a reinforced concrete slab together with buttress points to locate the jacks. The slab is held in position using ground anchors (see fig. 6.6). When the heading has passed through the complete area of exposed soil, the cutting edge of the leading section is removed. Raked flank walls are then constructed. The electricity supply to the street lights can be tapped into, in order to provide lighting for the heading (or culvert). The construction described is for a pedestrian subway. The procedure for a service duct is similar but a smaller-section circular pipe is used so that ring compression is produced. This may eliminate the need for reinforcement.

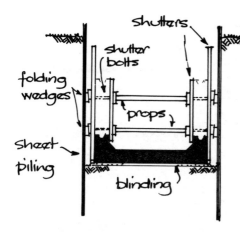

BASE STAGE :-
Sheet piles driven and excavation made. Base area blinded and base cast with joint to wall Kicker. Wall shuttering installed.

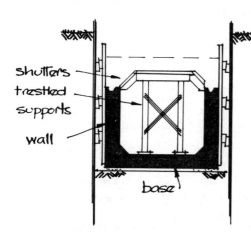

WALL STAGE :-
Walls cast with joint to roof section. Roof slab cast in short runs with shutters moved in alternate bays

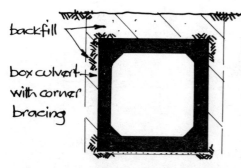

REINSTATEMENT STAGE :-
As roof slab shutters are removed, side shutters and sheet piling are withdrawn and space backfilled

Fig 6.5 Construction of Box Culvert by 'Cut and Cover' Technique

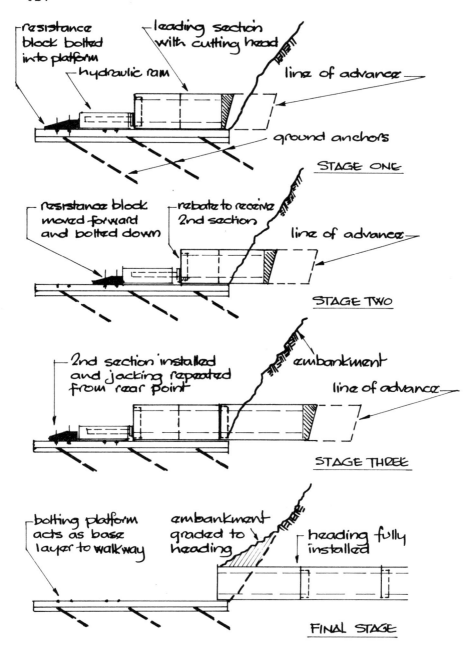

resistance
block bolted
into platform

leading section
with cutting head

hydraulic ram

line of advance

ground anchors

STAGE ONE

resistance block
moved forward
and bolted down

rebate to receive
2nd section

line of advance

STAGE TWO

2nd section installed
and jacking repeated
from rear point

embankment

line of advance

STAGE THREE

bolting platform
acts as base
layer to walkway

embankment
graded to
heading

heading fully
installed

FINAL STAGE

Fig 6.6 CONSTRUCTION OF BOX CULVERT
BY 'JACKED HEADING TECHNIQUE

Open culverts for canals and storm drains are constructed in a similar way to that described in 'cut and cover'. However, because they do not have a top section, their behaviour is different, resulting in the reinforcement being transposed to take tension on the opposite face of the wall section.

SIMPLE ROAD BRIDGES

A road bridge may be used to span over a stream, a railway, another road or a pedestrian area. At least two of these situations are such that the existing service cannot be disrupted without creating extreme economic and social problems. How then can we construct the bridge?

Initially we must construct the bridge abutments. These form the supporting element to the bridge deck and can be in one of three materials: brickwork, in situ reinforced concrete (r.c.), and steel sheet piling.

Brickwork is suitable only for very small bridges, having neither the strength nor the stability of reinforced concrete. Steel sheet piling is suitable mainly for river works, being easy to locate and fix. In situ r.c. is used mainly for abutments to bridges over roads, and forms retaining walls to each end of the new embankment. Counterfort walls may therefore be built, with the embankment work built up to the rear of the wall as backfill.

To avoid the use of any falsework or centring of temporary works that would be necessary with an in situ r.c. bridge deck, we will use prestressed bridge beams. These are standard beams of various spans and cross-sections. We will need a mobile crane for the installation of the beams and this should be phased to coincide with the delivery of the materials in the form of beams.

The beams are slung at $\frac{1}{5}$ points to minimise bending effect and are manoeuvred into their positions at the abutments. [*Note*: The seatings to the abutments should provide a sliding bearing to each unit to take account of expansion and contraction. P.T.F.E. (polytetrafluoroethylene) is ideal for this, having low frictional properties.] The movement joint at each end is masked at road surface level by an open saw-toothed fitting to the road and the bridge (see fig. 6.7). The infill between the prestressed beams is of cast in situ concrete, which helps key the beams and provides a base from which to develop the road finish. The surface of the bridge deck should match that of the new road, which will be determined by whether it is of flexible or rigid pavement construction.

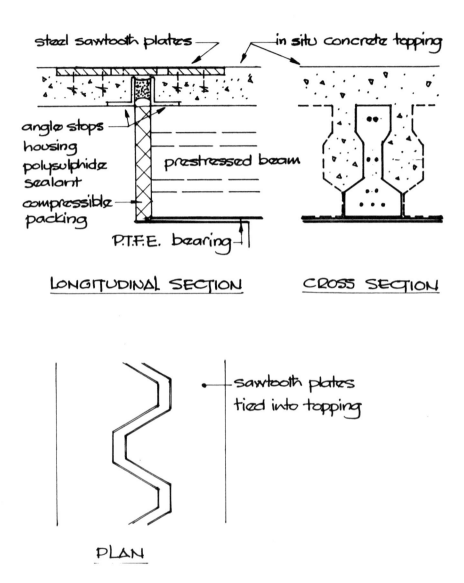

steel sawtooth plates

in situ concrete topping

angle stops
housing
polysulphide
sealant

compressible
packing

prestressed beam

P.T.F.E. bearing

LONGITUDINAL SECTION

CROSS SECTION

sawtooth plates
tied into topping

PLAN

Fig 6.7   DETAIL AT ABUTMENT BEARING
SHOWING TREATMENT OF MOVEMENT JOINT

ROAD CONSTRUCTION AND MAINTENANCE

The road construction described in chapter 7 of Volume 1 is substantially similar for country roads. The decision on whether to use flexible or rigid forms of construction will depend, to a large extent, on the nature of the subsoil. This in turn will affect the road surface selection and the problems of maintenance. The major differences by comparison with estate roads are:

(a) drainage,
(b) construction techniques,
(c) scale,
(d) maintenance.

Drainage of roads in open country can generally be achieved using land drains or ditches. If the road is slightly above the level of the fields or woodland through which it passes, the surface-water can be 'run off' into them and fed into their drainage system where appropriate. [*Note*: Where ground-water poses problems, a land drainage system of perforated pipework feeding into sand drains or deep soakaways may be installed by the landowner/farmer.] If, however, the road lies slightly below the general ground level, the simplest solution is to provide drainage ditches until suitable subsoils occur along the length of the road. Flexible roads in wooded country are still cambered to throw off rain and melting snow to both sides.

Trunk roads, because they may incorporate central reservations to separate the traffic, will more commonly fall to one side so that rain is cleared more quickly from the overtaking lane. Where a lay-by is provided, a short run of drainage channel with gridded protection may be installed, serving a secondary function of forming a visual divider. One very important factor to remember when roads are built into hillsides is to ensure that the fall is directed *towards* the hill. Although this will then necessitate some form of drainage gully, it will prevent surface-water running off the hillside from draining across the road. The logic behind this preventive measure is that such surface-water will generally contain some small earth particles. If this runs across the road it will produce a fine mud surface that promotes skidding. The direction of fall would then drift the vehicles towards the edge of the hill (see fig. 6.8).

Construction techniques for long runs of road in open country will involve a more 'plant-intensive' approach than that used for estate roads. Consolidation and soil stabilisation form an important part of

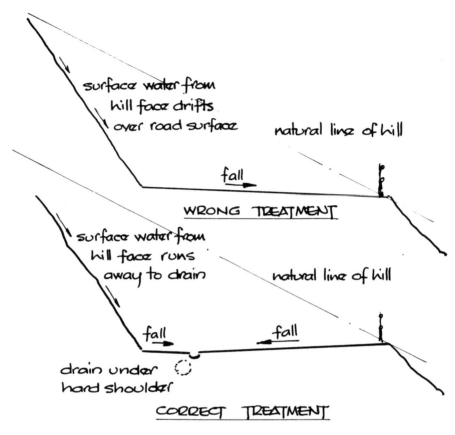

surface water from
hill face drifts
over road surface

natural line of hill

fall

WRONG TREATMENT

surface water from
hill face runs
away to drain

natural line of hill

fall

fall

drain under
hard shoulder

CORRECT TREATMENT

Fig 6.8 DETAIL OF BAD AND GOOD PRACTICE
FOR DRAINING ROAD CUT INTO HILLSIDE

the preliminary work, which is executed as described earlier in the chapter. The installation will then be carried out using a 'paving train' or, in the case of rigid roads, a 'slip form paver'. The paving train is a logical sequence of operating machinery linked together to ensure that the processes are carried out in a controlled sequence at consistent intervals. Each item of plant will serve a specific function, e.g. for a flexible road the processes in the train will include spreading and compaction of sub-base, installation of road base (including compaction), laying of base course, and spreading and finishing of wearing course. The sections of the 'train' are fed by lorries to ensure a continuous supply of the selected materials. Sighting wires will ensure the control of course thickness and cross fall.

The slip form paver, used for rigid road construction, condenses the processes into one composite machine rather than a collection of individual items. This is possible because the construction form involves fewer processes, e.g. concrete is discharged from lorries into the paver, which spreads and compacts it, feeding it into the slip form to establish its thickness. Expansion joint filler is fed into a cartridge and dowel bars into a magazine to form the expansion and contraction joints as required. Reinforcing mesh is laid in mat form from a roller into the 'plastic' concrete to the required depth and a profile board reinstates the surface. As with the paving train, sighting wires control cross fall. To the rear of the former a 'comb' produces a random ridged surface to assist drainage and reduce skid effect. A curing membrane is then sprayed on to the surface. Temporary joint fillers are plugged into the surface at movement joints for stripping at a later stage, the resultant gap being filled with polysulphide sealant.

Scale of road widths will vary with proposed usage, major trunk roads having up to three lanes in each direction whereas small 'B' roads may be narrower in places than estate roads, and widened at intervals to provide passing points. For this type of road it would be illogical to use a paving train, and traditional techniques are adopted.

Maintenance is the major consideration on all open roads, as can be seen by the constant scatter of road works across the country. In Volume 1 we looked at ways of reducing this aspect of the roadworks by logical location of services. This can be developed further by containing services in culverts where they cross under roads. Correct stabilisation, compaction and grading of the sub-grade is essential to avoid settlement of the surface. Even with all these precautions taken into account, the continuous traffic of heavy lorries will wear the road surface.

Surface repair work usually involves 'scoring' the concrete or wearing course. The purpose of this is two-fold: first, it reduces the degree of uneven wear, and secondly it provides a key for the resurfacing. Certain precautions should be adopted in the execution of this resurfacing. These are:

(a) Ensure that any grids, gullies or drain inlets are protected during the scoring and resurfacing to avoid damage or blockage.

(b) Ensure that the cross fall or camber is retained to accord with the original specification.

(c) Check that 'ponding' does not occur, creating pools of surface-water.

These sound obvious but they are frequently overlooked owing to poor quality control.

Another aspect of maintenance is regular inspection of land drains to check their functional efficiency, and, in the case of small country roads, clearance of drainage ditches. Associated work that is not in the strict sense maintenance is road clearance. This is mainly seasonal, for example autumn will involve clearing leaves and small branches; winter may include gritting and possibly snow clearance; spring will involve verge clearance and hedge trimming of small country lanes.

Whichever aspect of external works we consider, it is important to realise that more people see and use these elements of construction than any of the forms described in other chapters. If things go wrong they will affect more people and faults will be more noticeable. This means that the level of supervision required is greater and more varied than for more specialist works. This is the province of the technician and it is vital that the student should develop a full understanding of problems and developing techniques. This is best done by reading journals of an international nature on ground engineering and civil engineering construction as a method of continual updating.

Assignment 6.1

Select a large area of open countryside, possibly from an ordnance survey map, and obtain a geological map of the area. Plot the route of a small road to avoid natural hazards where possible and identify zones that require embankments, cuttings or small bridges. Based on topographical and geological data, produce a reasoned argument for the type of pavement that should be adopted.

Assignment 6.2

Produce shuttering details for a 20 m length of counterforted reinforced concrete retaining wall, allowing provision for the location of a land drain at the junction of base and stem. The wall is to receive backfill on completion. [*Note*: It would be helpful in this assignment to refer to brochures produced by specialist shuttering and falsework manufacturers.]

7 Execution of a Contract

In this chapter we shall try to include a number of the forms and techniques of construction discussed in the earlier chapters to show interrelationships, where they occur, and the need for logical planning of every facet of the contract. It is not intended to include contract administration or management in any detail, since this side of the construction process is more realistically considered at higher award level.

What we shall try to achieve is a development of logical thinking and an ability to look ahead to problems that might occur so that they can be allowed for in the overall time schedule. There is nothing worse than working under the pressure of an unrealistic deadline that cannot be met regardless of incentive schemes, bonus payments and overtime working.

Leaving aside what the actual contract comprises, let us examine some features that are common to most contracts. Who is involved in the design, specification, costing, construction, installation, control and supervision?

DESIGN

The client will have an interest in the design aspects of the work since it is he who is paying for it. The prime responsibility, however, will rest with the 'design leader'. This role will usually be taken by an architect or a building surveyor, since they integrate the various design elements to produce the general arrangement drawings that form part of the contract. In some types of contract this is not the case, the role being adopted by a quantity surveyor with a remit to keep a tight control on the overall costs of the contract. In other types of contract where the work is of a specialist civil engineering nature, such as a major road bridge or deep-water harbour, a structural or civil engineer would lead the design team. For the majority of contracts, however, even if a quantity surveyor leads the design, its actual execution will rest with the offices of an architect or surveyor, a structural engineer and a mechanical/electrical services engineer.

The architect's (surveyor's) office will use a small team to run its aspect of the design. This might typically comprise:

one chartered architect (overall control)
two or three higher technicians (design aspects)
three or four junior technicians (detail aspects)
one or two tracers (revision work/copying)

While contact might be through both the architect *and* the higher technicians, it will certainly not drop below that level when related to discussions with engineers or contractors. The client will probably deal only with the architect.

The consulting engineers' offices, both structural and services, will probably have a similar structure such as:

one chartered engineer (overall responsibility)
two technician engineers (design responsibility)
two higher technicians (design/detail aspects)
two junior technicians (detail/schedule aspects)
one tracer (plotting of other drawings/revisions)

Contact in this case will generally be through the technician engineer and occasionally a specified higher technician.

SPECIFICATION

The client will also have a keen interest in the specification of the work but the prime responsibility will rest with the design leader. Depending on who this is, the other people involved in the drafting stage will be:

architect's office − one higher technician
structural office − one technician engineer
services office − one technician engineer
Q.S. office − one quantity surveyor

Additional involvement at a later stage will be from most of the personnel listed above, plus the contractor's quantity surveyor and the contractor himself. It might also include:

quantity surveying technicians
site agent/manager

senior site technicians
building control/clerk of works

COSTING

Costing is the province of the quantity surveyor or, more realistically, senior technicians within the practice, on both the client's and the contractor's side. It will also involve the technician engineers for the structural and services work. This is because they will be more familiar with the type of work to be costed and can offer considered opinions on the reliability of quotes from subcontractors tendering for work.

CONSTRUCTION

The responsibility for this work, even if subcontracted, will rest with the main contractor, the exception to the rule being when 'nominated' subcontractors are used. As with the other organisations, a structure of command is set up. In this case, however, it is somewhat larger, comprising:

one contracts manager (overall control)
one site agent (depending on the firm)
one general foreman (liaison with trades)
two senior technicians (setting out/general supervision)
specific trades foremen (control of trades)
associated craftsmen (production of work)
semi-skilled labour (support role of work)
labourers (heavy-duty work)

The associated trades will often include:

carpentry and joinery
brick- and blocklaying
plastering and rendering
plumbing and gas central heating
electrical and H.V.A.C.
roadworkers
asphalting/roofing
landscaping

All contact is through the respective trades foremen and, with respect to discussions with the architect's, engineer's and quantity surveyor's offices, will be through the general foreman, who might involve the trades foremen and senior technicians in the discussion.

The discussion will take one of five forms, namely:

letters – correspondence
telephone
written instructions
site visits – inspection or interim valuations
site meetings – progress, variations

Accurate records should be kept of all discussions and these will vary, for example correspondence can be filed, telephone conversations should be summarised in note form, with the main decisions formalised by letter. Written instructions can also be filed but recording of site visits will depend on the nature of the discussion. Agreements to modifications will need formalising by letter, acceptance of valuations or confirmation of quality control test results may need signature. Site meetings should always be recorded, either as minutes or as a summary of the points discussed.

INSTALLATION

Subcontractors nominated to install foundations or mechanical services, etc. will have structures of command similar to that of the main contractor but these will involve a much smaller team. Liaison will be through the main contractor, who may involve the consultant's office in technical discussions. The main contractor's senior technicians might well become involved in such liaison.

CONTROL

Although this is primarily the concern of the local or district authority through the building control office, district surveyor's department or the clerk of works, it will also involve internal control. The designer may get involved and would afford responsibility for this to one of his higher technicians. The consultants might install a 'resident engineer', usually a young graduate or technician engineer, but might use the system of site visits as sufficient control. The quantity surveyor will also exercise indirect control of the quality of work in agreeing (or not) to issue stage payments. Above all, the contractor will operate his own quality control to reduce the risk of delays caused by remedial work. Time and materials lost for rejected work are his direct responsibility.

SUPERVISION

Supervision is sometimes confused with control but although it tends to encompass quality control, particularly in the testing of concrete cubes for approved strength, it is concerned very much with site safety, security and working conditions. It is also concerned with working techniques, the evaluation of those that have been introduced recently, and the assessment of staff capabilities. Knowledge of this last factor is of vital importance when assessing time scales for construction during the estimate stage. An appreciation of time and equipment requirements for specific jobs will help in planning the network of items in the contract, possible variations in start and finish times and any specific equipment that may need to be hired on short-term contract. This area of study is also developed more fully at higher award level and provides a good foundation for the study of management techniques. Examples of its application will be shown in chapter 8 in relation to the scale of work covered at this level of study.

Let us now illustrate, by application, the features discussed, with reference to a typical case study involving a small amount of external works, some low-rise construction and some subcontracting. We will take a general look at the overall contract with a more detailed examination of some areas of the work.

CASE STUDY 7.1

A small site, once occupied by a cinema, is to be redeveloped to house a supermarket, with offices over, and a multi-storey car-park. The supermarket/office block is to be of steel-framed construction to a 7.5 m square grid. The floor slabs are to be of hollow ribbed in situ reinforced concrete supported by castellated steel beams in both directions. Foundations are to be based 2.0 m below street level and the standing ground-water table is at 1.5 m below street level. The car-park is to comprise two blocks linked by access and egress ramps. Its construction is reinforced concrete cantilever frames supporting one-way hollow ribbed floors. The storey height for each block is 2.8 m, giving a ramp inclination of 1 in 5. All surfaces are to be self finished, the columns are to be provided with corner buffers, and the floors are to be laid to falls. The car park is open with perimeter upstand beams provided to 'contain' cars. The external finish of these beams is of striated concrete and is achieved using precast panels as permanent formwork. Figure 7.1 shows plans and general arrangements for the development.

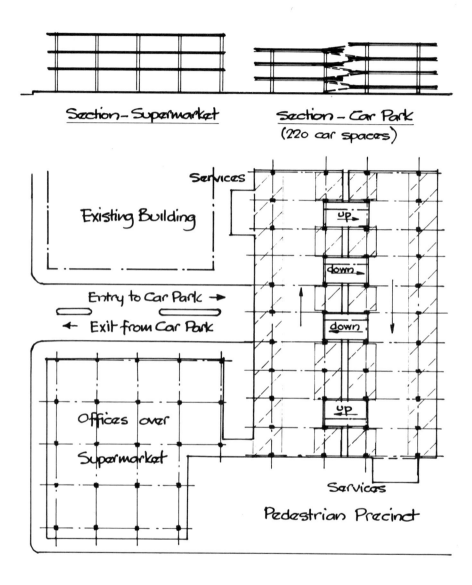

Section - Supermarket

Section - Car Park
(220 car spaces)

Services

Existing Building

Entry to Car Park →

← Exit from Car Park

up

down

down

up

Offices over

Supermarket

Services

Pedestrian Precinct

Fig 7.1 GENERAL ARRANGEMENT DRAWINGS
TO SUPERMARKET/OFFICE COMPLEX CONSIDERED
IN CASE STUDY 7.1

Reasons for the design

The designer, in consultation with the structural engineer, chose steel-framed construction for the main building because it offered the ability to provide a reasonable column grid. It was important that a fairly 'open' layout was made possible for circulation space within the supermarket. This allowed for palleted merchandise to be moved around easily when stocking out shelves. The layout at office level was not so critical but benefited from the generous column grid. The choice of hollow ribbed r.c. floors supported by castellated beams was to minimise the dead loads in two ways:

(a) the castellated beams are more efficient for their weight than solid steel;

(b) the ribbed floors keep concrete volume low and by spanning both ways provide better resistance to deflection.

The selection of an in situ r.c. building for the car-park was made on the basis of fire protection, durability in case of impact and the ability to produce a building that needed no further finishes to be applied — steelwork would have needed 'solid' fire protection so shuttering would have been necessary for the beams and columns.

Having made the decisions regarding the overall structural form, what claddings should be provided? The choice of striated concrete for the car-park was straightforward. Since all the interior was concrete, the outside should match it and the best weathering characteristics were thought to be vertical striations. Should this be repeated for the supermarket/office block? Visually it made sense that the two buildings should relate. Also nobody would see the steel frame because of its fire protection. But how could the panels be restrained, and how could windows be fixed? The method chosen was to introduce concrete mullions (precast at the same works that produced the panels), which would run the height of the two storeys of offices. These would tie into the floor slabs and would break the panels down to 2.5 m lengths tying in with the car-park module. Unlike the car-park units, however, these panels would *not* be acting as permanent formwork but would include insulation, sandwiched between outer and inner faces, and window-cills and heads (see fig. 7.2). Externally they would have the same appearance, and to keep thickness to a minimum it was decided to use stainless steel reinforcement of small diameter.

What floor finish should be provided to (a) the supermarket, and (b) the offices? The supermarket floor would be sustaining heavy

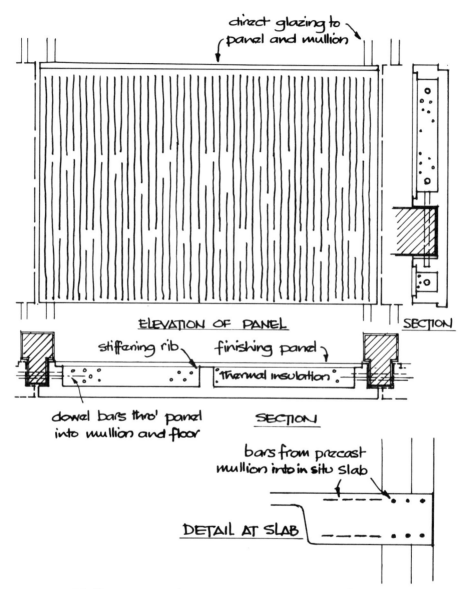

Fig 7.2   DETAIL OF PRECAST   CLADDING PANEL TO SUPERMARKET/OFFICE BLOCK CONSIDERED IN CASE STUDY 7.1

loads, would be subject to abrasion and, because of the heavy pedestrian traffic and probable breakages of bottles and jars, would need to be easily cleanable. Polished granolithic paving slabs were chosen with polysulphide joints to provide an impervious surface of proven durability. The office floor finish needed to have some sound-deadening characteristics while allowing for installation of communication and electrical services. A floating finish of 60 mm polystyrene slabs topped with tongued and grooved chipboard was selected, primarily because it made service location easy, secondly because the chipboard presented a good fixing surface for carpet tiles, and thirdly, by way of a bonus, because the resultant thickness masked the support nib of the precast panels.

Since an air-conditioning system was considered essential for the supermarket it was decided that it could be extended to service the office floors as well. This meant that a suspended ceiling should be provided, incorporating 'vent' grilles, lighting panels and (for the supermarket) speakers for communications and background music.

Insulation at roof level was considered, and lightweight bituminised aggregate selected for the reasons outlined in chapter 3 under 'inverted insulation construction'. The only remaining consideration was the type of windows/glazing. Because the offices were to be air-conditioned, fixed-light double glazing was chosen, this being reflected at ground level by sealed storey-height double glazing units and automatic opening double glazed doors. The glass was to be solar tinted and (in the case of the supermarket) toughened against impact damage.

Major points regarding the specification

At this stage we do not go into the writing of specification clauses and we will content ourselves with an outline of the more obvious items of work.

Construction materials will generally need to meet either B.S.S. requirements, or C.P. strengths in the case of the reinforced concrete. Those not covered by either of these will have to be Agrément certificated. The concrete will be specified to cover strength, workability and aggregate size. Why do we need to specify these *three* requirements? Strength is obvious — it has to be strong enough to perform adequately. Workability is less obvious. It must be possible to place the concrete in the shutters provided. For columns this will require *high* workability (or medium for very large cross-sections). Whereas solid slabs only require low workability, the ribbed floors make concrete placing more difficult and so *medium* workability is specified. This is also the case for the car-park upstand beams. In addition, aggregate size is important, since it must pass between the reinforcing bars as well as around them at the shutter face. Normally, 20 mm aggregate (this being the maximum size) would be perfectly acceptable, but in considering the precast cladding and the ribs of the floors it would be sensible to specify 12 mm to allow for easier placing. The use of stainless steel, for the cladding panels only, would also need to be specified.

Equipment such as the air-conditioning plant would have to meet performance specifications and be covered by guarantees. The testing and commissioning of the system would also be written in. This would apply equally well to the communications systems and electrical supplies.

Quality of workmanship would be specified and if, when inspected, it did not meet the specification, the contractor would have to take remedial action. This might (in a very serious case) mean demolishing the unsatisfactory work and producing new items. The contractor would therefore incur additional expense, which he would have to offset against his profits, so it is always in his interests to 'meet the specification'.

Quality of finish is also important, since poor quality control in this aspect of the construction can mar the appearance of the work and possibly fail to give the correct degree of protection to the structure. It might even affect the functional performance of the building. For examples, in the case now being considered, the floor surface of the car-park must provide for drainage, assisted by the tamped surface, and the surface of the ramps must be skid resistant.

This list is by no means exhaustive but is intended to give an appreciation of the need to agree in writing the type of quality on which the contract may be priced. This will affect the contractor also with respect to his choice of labour for the contract.

Reasons for the costing
This is necessary in order to establish, quite firmly, the costs to which the client or developer will be put. Since *time is money* it will also involve a time limit to be observed. If you ever read the book *The Forsyte Saga*, or saw the T.V. serial, you will recall the problems of cost that arose because the architect was not bound either by materials and labour costs or by a time schedule. A penalty clause based on completion date will almost certainly have been written

into the contract for the present project because the launch date for opening the supermarket will have been planned fairly precisely and any delay will result in loss of takings. Because the date is so critical in this instance, an incentive bonus will have been written in to encourage early completion.

The materials, components and equipment for any contract need to be costed by volume or item so that *stage payments* can be made. This involves a process known as *interim valuations*, where the quantity surveyor, acting for the client, visits the site at agreed intervals to inspect the work that has been completed and the unused materials that have been purchased. When he is satisfied as to the quantity of work completed and materials/equipment stored on site (and this may involve discussion with the contractor's quantity surveyor), he will agree to the payment for work done. This is primarily in the interests of the contractor, since he is often working on borrowed capital, and it is important that his cost control is effective to maintain a steady flow of stage payments.

Also built into the costing is *retention money*. The retention clause is written in to cover the initial maintenance period and requires the making good of any defects (such as faulty air-conditioning, poor finishing causing surface deterioration, or ineffective cladding causing rain penetration) that occur after the buildings have been made operational. The retention money is held by the client to cover such making good and he can refuse to pay this money until satisfied that the building meets the specification agreed in the contract documents.

Construction considerations

The contractor's initial considerations in respect of the construction will be made at the tendering stage. An examination of the specification, bill of quantities and contract drawings will have enabled him to determine his approach to planning the construction prior to pricing and timing the overall contract. This might, for example, include some subcontracting of work such as the steelwork erection. This would be logical since it would otherwise mean the use of specialist labour for only a part of the contract.

A second area of preliminary consideration is a 'pre-contract' discussion with the designer, structural engineer and quantity surveyor and possibly the services consultant. This aspect of the contract is very useful since some minor modifications, agreed at this stage, might help to streamline the work. It also helps to establish an understanding between all the interested parties in setting up lines of communication. An example of the sort of discussion would be how best to deal with the foundation work. Because the ground-water level will mean setting up a 'well point' system as a temporary de-watering process, it would be logical to produce all the work below the standing water level at the same time rather than treat each building independently. It would probably involve a decision to nominate a mutually acceptable supplier for the precast concrete cladding, thus enabling the delivery schedule for such work to be planned.

Dewatering of the site will necessitate an analysis of the site investi-gation report, particularly the borehole logs, the gauge the volume of ground-water likely to be handled. An agreed disposal method for this will also need to be approved. A second aspect of the site investi-gation is the analysis of the soil and ground-water for sulphate content, to be confirmed once the work gets under way. Should a change of concrete specification be necessary, all interested parties will have been fully prepared. This should reduce the risk of delays caused by arguments about correct procedures for such change.

On the basis of these preliminary considerations the contractor can now analyse the work involved. This is done by compiling a *method statement* for each component part of the contract, including subcontract work both 'nominated' (i.e. that of subcontractors selected directly by the client for specific elements of the contract) and that for which the main contractor is fully responsible. Having estimated the materials, equipment and labour requirements for each component, together with a time allowance, the contractor can translate this into a *precedence* chart. This means identifying what work *must* be done before other work can start. Some areas of the work will be more open to variation than others, which means that flexibility can be built into an overall network of events.

The main benefits of this pre-planning are:

(a) A logical programme of activities, with optimum use of labour, can be determined.
(b) Phasing of materials/component deliveries will minimise site storage and double handling, which in turn will reduce the risk of damage and pilferage.
(c) Timing of subcontract areas of the contract can be more accurately assessed.
(d) Sensible stage payment times can be identified.
(e) A *critical path* of activities can be programmed, showing

alternative routes if problems should occur in some of the work phases.

This work is dealt with more fully in higher award studies but figs 7.3 and 7.4 give a simplified example of the way in which these systems operate.

Construction details

There are some aspects of the construction dealt with in this case study that we have not covered in previous work and it would be sensible to consider them in some detail. This will involve examining both how a particular technique is carried out and why its detail is important.

The *steel frame* utilises castellated sections on a two-way spanning frame. The seating arrangement of beam to column is basically similar to that shown in Volume 1 but involves heavier seating angles to accommodate the extra load. Also, because the beams are expanded, they are less resistant to shear at their support areas and need to be stiffened by web plates to resist the loading. Details are shown in fig. 7.5.

The *two-way hollow ribbed flooring* spans between and across the steel frame with some curtailment of the reinforcing bars to use the steel reinforcement economically. Also the area around the column is reinforced diagonally to provide resistance to 'racking' (torsion). The 'pots' are produced using polypropylene moulds, which are removed on completion by means of compressed air. Details are shown in fig. 7.6.

The *concrete cantilever frame* of the car-park is reinforced in its tension faces and has a concentration of shear reinforcement towards its supports. The hollow ribbed floor, spanning only in one direction, has the ribs staggered at supports to give a more effective compression zone and to avoid too sudden a change of cross-section. Details are shown in fig. 7.7.

The *ramps*, which connect the two wings of the car-park, are supported by nibs formed in solid bands of concrete along the column zone. The detail is such that a simply supported bearing is formed at both limits to allow some rotation in the event of minor differences in settlement occurring between the two wings. Details are shown in fig. 7.8.

The precast concrete used as *permanent formwork* is tied into the in situ upstand walls and comprises two parts, the cill member being

SUPERMARKET/OFFICES   CAR PARK BLOCK

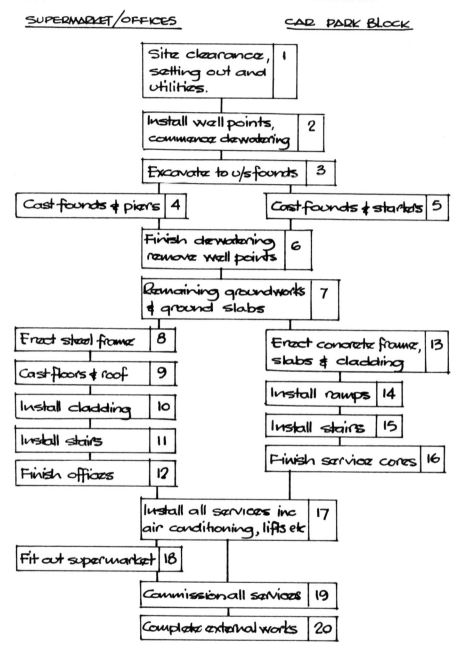

Fig 7.3   PRECEDENCE CHART FOR

CASE STUDY 7.1

KEY :—

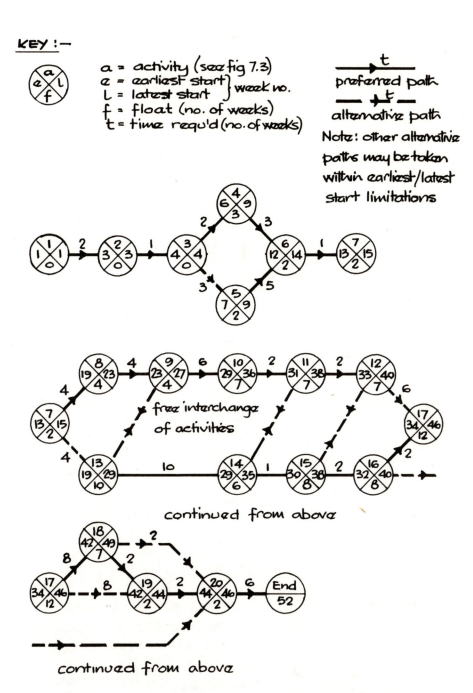

continued from above

continued from above

Fig 7.4 CRITICAL PATH LAYOUT OF CONSTRUCTION ACTIVITIES FOR CASE STUDY 7.1 (read in conjunction with fig 7.3)

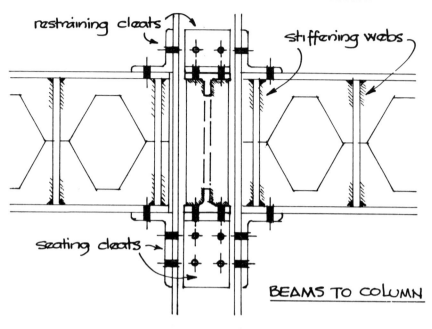

restraining cleats

stiffening webs

seating cleats

BEAMS TO COLUMN

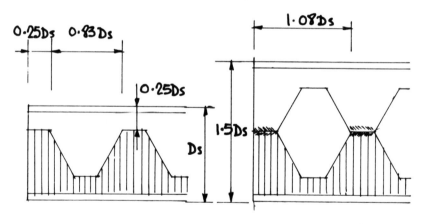

0.25Ds   0.83Ds

0.25Ds

1.08Ds

Ds

1.5Ds

beam cut to profile
using 60° castellation

top section lifted,
transposed and welded
to bottom section

Fig 7.5 DETAILS OF BEAM/COLUMN JOINT
FOR SUPERMARKET/OFFICE IN CASE STUDY 7.1
AND SKETCH SHOWING BEAM FABRICATION

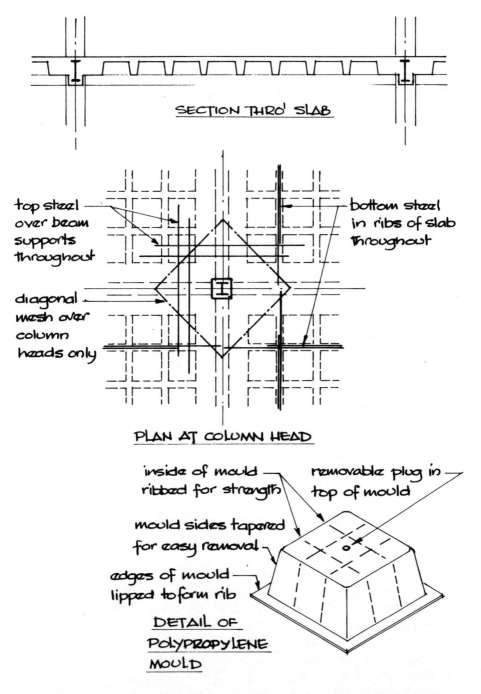

SECTION THRO' SLAB

top steel over beam supports throughout

bottom steel in ribs of slab throughout

diagonal mesh over column heads only

PLAN AT COLUMN HEAD

inside of mould ribbed for strength

removable plug in top of mould

mould sides tapered for easy removal

edges of mould lipped to form rib

DETAIL OF POLYPROPYLENE MOULD

Fig 7.6 DETAILS OF TWO-WAY RIBBED FLOOR SLAB TO SUPERMARKET/OFFICE BLOCK

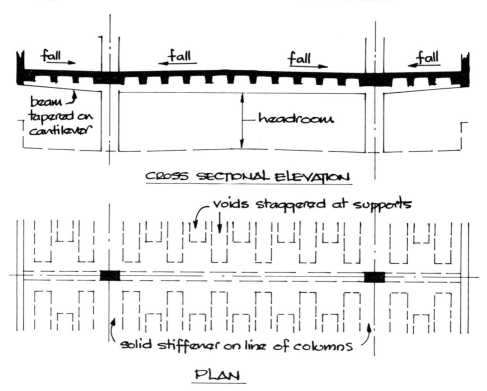

fall   fall   fall   fall

beam tapered on cantilever

headroom

CROSS SECTIONAL ELEVATION

voids staggered at supports

solid stiffener on line of columns

PLAN

Fig 7.7 DETAILS OF CANTILEVER FRAME TO CAR PARK BLOCK IN CASE STUDY 7.1

placed after the concrete is poured, to give a consistent external appearance. Details are shown in fig. 7.9.

Service installations consideration

Whereas the horizontal location of the services to the supermarket/ office building can be provided for in the floating floor finish and suspended ceiling space, any vertical transition must be allowed for by holes and/or service ducts through the floors. Obviously, this provision needs to be established before the start of the construction to avoid serious structural problems. The normal procedure is for the structural engineer to send a set of floor layout drawings to the services engineer for 'marking up'. This may result in further discussion as to duct location if the requirements pose serious structural problems. It

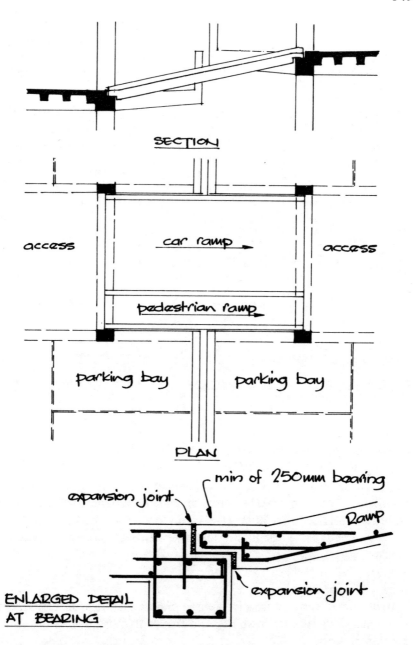

SECTION

access   car ramp →   access

pedestrian ramp →

parking bay   parking bay

PLAN

min of 250mm bearing

expansion joint

Ramp

expansion joint

ENLARGED DETAIL
AT BEARING

Fig 7.8 DETAILS OF RAMP BETWEEN
WINGS OF CAR PARK TO CASE STUDY 7.1

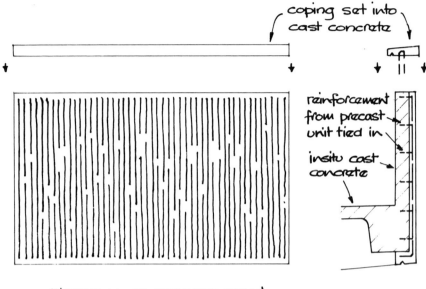

coping set into cast concrete

reinforcement from precast unit tied in

insitu cast concrete

ELEVATION OF PRECAST PANEL

Fig 7.9 PRECAST PANEL AS FINISH AND PERMANENT FORMWORK TO UPSTAND WALL

is essential that the final decision be reached sufficiently early to avoid delays. In the event of either no response or no decision being reached, the structural engineer, in consultation with the designer, will make provision on the basis of past experience, which might inhibit the services specialist. It is rarely necessary for this condition to be reached since its likelihood is incentive enough to reach a solution.

Process of control
Control of the contract falls into three main categories: materials and workmanship; structural stability and fire protection; and satisfactory progress of work.

The first two categories involve all parties plus the building control/ district surveyor's office in ensuring that the work produced conforms with that specified. This is sometimes complicated by the introduction of *variations* to the contract specification, which require the agreement of the design leader. Site inspections and visits normally form the basis by which this control is carried out.

The third category will not normally involve statutory controllers but will rather be confined to the construction team. It will generally take the form of site meetings at which the relevant parties are represented. These meetings help to identify any possible problems before they arise so that they can be resolved quickly and avoid costly delays to the construction process. The design representative is likely to be present at all such meetings but other parties might attend only part of a meeting if the query is specific.

The discussion and agreement of all points considered is recorded so that no decisions reached are ignored. Each representative is then provided with a copy of the report so that action, where necessary, can be taken. In the early stages of the contract the meetings will take place at specific times, such as at the completion of foundation work, to alert all concerned to the possibility of delays in specific areas. This process is aligned with the critical path to point out logical reshuffling of the programme to maintain the target date.

Assignment 7.1
Produce a method statement for the roadworks associated with the case study, i.e. approach road to the car-park, pedestrian areas around supermarket and offices. Explain reasons for the choice of construction to be adopted and sketch details of the road construction, showing provision for drainage.

Assignment 7.2
Suggest a suitable form of construction for the fire escape stairs to the car-park and sketch details of the construction and associated temporary works.

Assignment 7.3
Visit some multi-storey car-parks close to where you live and compare their forms of construction with the example covered in the case study. Examine, in particular, any applied finishes and assess their contribution to the overall efficiency of the buildings. Produce a brief critical analysis of the various forms of construction examined.

8 Site Supervision

By now we have seen that construction forms, techniques and sites vary to a very large degree, even for low-rise construction. This is what makes this branch of technology so different from those that are factory based. Technologies such as car production or steel production are highly specific and are made even more specialist in their component parts by a tradition of job demarcation.

A construction site could not run efficiently if designed along such lines and the various job specifications are only loosely tied by general disciplines. A bricklayer, for example, will work below or above ground to produce either facing work or work to be hidden by applied finishes. He may also turn to vertical tile hanging or roof tiling, and could be involved in the setting-out process. Because of this variety of job functions, which holds true equally well for a carpenter or a technician at a different level of operation, it is not sensible to operate the same principles that are adopted in factory work.

Time-and-motion study techniques using the 'clipboard and stopwatch' approach become nonsensical if applied to bricklaying. For example, a straight run of one-and-a-half-brick wall using flettons and operating from ground level would be vastly quicker to construct than a piece of half-brick facing work for a cavity wall with internal and external corners plus openings for windows and operating from a scaffold. An assessment of capabilities must be much more subjective and will usually be based on the performance of a team, for example two bricklayers and one labourer.

A craft foreman will have a fairly good idea of what can or cannot be done, and the time necessary, for his own particular craft based on his own experience. He can translate this to the performance of the craftsmen and labourers working for him, either upgrading or downgrading the particular capabilities of individuals to determine realistic time schedules for any job within his field of work. This will be based on a normal speed of operation with built-in quality control and *not* work under pressure. The latter *can* be sustained for short periods and is a useful asset when extreme weather conditions have produced delays but if applied for long periods it would probably result in a decline in quality control.

In much the same way a general foreman should be able to assess the ability of the various crafts foremen to control their personnel. By operating informal progress meetings or just discussing progress with the individuals, he will be able to measure their efficiency. The technician needs to become involved in this assessment process and should observe the role and capabilities of the various site personnel, making a *mental* record of time scale related to job difficulty and weather conditions. This becomes particularly important with the introduction of 'new' materials and techniques and is a continual updating process.

Only when a realistic appraisal of ability has been gained is it possible to *supervise* site personnel and production. This must, of course, be coupled with an understanding of the materials used in each process, their limitations and performance and the qualities that they should possess. It is in the development of understanding different construction techniques, and allying them to labour and material capabilities, that the technician becomes a key figure in site supervision. To communicate effectively with the various personnel involved he must know what he is talking about and must show this without upsetting people or causing unrest. If the craftsmen sense that a technician is unsure of what is being done or is 'waffling' by using pseudo-technical language, they will disregard his role. Similarly, if a 'supervisor' is officious and dogmatic about what should be done, he is liable to produce poor co-operation.

Production schedules and quality control are not the only aspects of site supervision. An equally important factor is that of safety. The construction site could, if allowed to do so, become the most dangerous work situation imaginable. We have already seen the dangers of working in excavated ground or with heavy moving machinery. If we recall the problems concerned with scaffolding, hoardings and hoists, we can see how associated work here needs to be supervised. These are the more obvious safety-conscious areas of site work but even simple items, like offcuts of timber with nails sticking up, or broken glass lying half-hidden in the mud, can produce serious accidents. In truth, accidents very rarely *happen* on a building site: they are generally *caused* by carelessness and untidiness.

An untidy site promotes a careless attitude and makes it difficult to spot the hazards that should be obvious, whereas a tidy site promotes a methodical attitude and a more safety-conscious environment. This methodical approach can be developed in looking at working conditions. It is important that hutting provisions should be sufficient to the needs of the site personnel but it is equally important that the other components of working conditions should be fully understood. Operatives working in dangerous or extremely uncomfortable conditions should not be expected to work for long periods without adequate breaks. Where possible, work should include some variety to avoid monotony, boredom and the risk of complacency. Overtime should not become the rule but rather should be used to bring a contract back on to schedule where delays have occurred, for example where a component manufacturer has been late with deliveries.

Lastly, let us re-examine the peripheral areas of the construction process: security, adjacent properties and the immediate effect of the construction on the local environment.

Security means safety from unlawful access, the control of stock to avoid petty theft, safe storage of materials and components, and provision against accidents from trespassers.

Adjacent properties must be protected from the effects of construction, e.g. noise, dust, rubble and smoke. This may involve the provision of tarpaulins to runs of scaffolding; 'mufflers' for pneumatic drills and 'fans' to hoardings. It must also be realised that when excavating close to an existing building its foundations could be undermined. If this were to happen, the contractor would be liable to substantial damages and it is important to establish the structural condition of adjacent property subject to such risk. This area of work is dealt with at higher award level and will often involve the use of shoring and underpinning.

The local environment can suffer in several ways because of the effects of a construction site. The dust, particularly from cement, might coat the leaves of trees and prevent them from functioning properly, resulting in serious damage. The mud from lorries leaving the site, if spread across the highway, could cause problems with roadholding for the general traffic. Litter blowing around, such as empty cement bags or polythene sheet wrapping, will create an unsightly environment. This should not be allowed to happen and part of the supervisory role is to protect against this. For example, light mesh reinforced polythene sheet fixed to a scaffold screen will protect trees from dust; a lorry grid located at the site entrance will clear most of the site mud from the tyres and a check on the spoil load will reduce the risk of spillage; control over the litter on site, an extension of the theme of a tidy site, will minimise the problem of pollution to the local environment.

Having talked about the supervisory aspect in general, let us develop some components of a building project to illustrate its implementation more clearly. This can best be done by examining a case study.

CASE STUDY 8.1

A small development of six houses is to be built on a plot previously occupied by two large Victorian houses. The site fronts on to a bus route and has all main services. It has already been cleared ready for new construction work and detailed planning permission and Building Regulations clearance have been granted. The form of construction is to be cavity-walled brickwork with trussed rafter roofing. We shall now examine some of the component parts of the construction programme with specific reference to site supervision.

Items for supervision

The list that follows is not meant to be exhaustive but to show areas of responsibility from within and outside the contractor's organisation.

Contractor	*Design Team*	*Building Control*
(1) *Initial setting out*		
(a) check on levels related to datum and existing drainage runs	check for accuracy	
(b) check on accuracy of building outlines	check for overall dimensions	check that boundary distances are correct
(c) check on correct transfer of levels to profile boards and sight rails		
(2) *Excavation work*		
(a) check for correct depth of all trenches	check on quality of sub-grade	inspection of sub-grade for foundations
(b) check for falls to drainage runs and gravel bedding		check for thickness of gravel
(c) check on stripped ground		check for topsoil removal

Items for supervision (continued)

Contractor	Design Team	Building Control
(3) *Casting of foundations*		
(a) control on concrete mix, including water content	check for quality of materials	sampling if thought suspect
(b) accuracy for thickness of footings and their level		inspection of footings
(c) check for bedding of inspection chamber pre-formed base units	check for quality of work	
(4) *Initial brickwork courses*		
(a) check for consistency of cavity		
(b) check on setting-out of walls from profile lines	check for accuracy	
(c) establishment of bond to minimise use of cut bricks		
(5) *Installation of drainage runs (soil-water)*		
(a) visual check on straight-ness of runs and correct jointing	check on layout to conform with design drawings	
(b) check on 'building-up' of inspection chamber walls	visual check for quality	
(c) check where drains are located inside buildings	check for location to accord with plans	
(d) check on provision of lintels as brickwork built around drainage pipes		check for lintel supports
(e) initial test for water – tightness of system prior to backfill		
(f) check on correct layer grading of backfill		inspection of backfill and test system
(6) *Installation of drainage runs (surface-water)*		
(a) similar checks to (5) plus location of rain-water gullies to houses and road	check for adequate protection to gullies	inspection of system

Contractor	Design Team	Building Control

(7) Brickwork up to and including d.p.c.

Contractor	Design Team	Building Control
(a) check on consistency for line and level	visual quality check	
(b) check on lapping of d.p.c.	visual quality check	inspection

(8) Hardcore and blinding to houses and road

Contractor	Design Team	Building Control
(a) initial check on hardcore to be used	visual material check	
(b) check on correct placing and compaction to houses		inspection
(c) check as (b) for roads, with emphasis on compaction	visual inspection for weak areas	inspection
(d) check on gullies	check against damage	
(e) check on thicknesses of hardcore and blinding	visual quality check	

(9) Oversite concrete, d.p.m.s, aprons and driveways

Contractor	Design Team	Building Control
(a) quality check, particularly if site mixed		
(b) correct installation, lapping and quality of d.p.m.s	visual check	
(c) check on placing time for concrete — before initial set — and provision for services	visual check on placed concrete and sleeving for gas, water, etc.	inspection of concrete work
(d) check on joint provision for driveways		
(e) check on work around gullies and inspection chambers, including fitting of covers to latter	check on quality of work, surface finish and possible damage to fittings	

(10) Cavity walls up to first-floor level

Contractor	Design Team	Building Control
(a) check on quality of brickwork, blockwork, wall ties and insulation board	visual inspection of materials	

Items for supervision (continued)

Contractor	*Design Team*	*Building Control*
(b) check on mortar strength and consistency		
(c) check on 'building-in' of insulation board to retain 25 mm cavity		inspection for walls to meet U value
(d) check on location of wall ties		inspection for spacing of wall ties
(e) check on verticality and quality of construction	visual check on quality of work	check for verticality
(f) check on details around openings (including lintels and d.p.c.s) and tying-in of frames	visual check on work to to accord with drawings	inspection of d.p.c. locations and lintels used
(g) check on inner load-bearing and partition walls	check to confirm locations to accord with design drawings	check for tying-in to external walls to provide lateral restraint
(h) check on stability of staging and protection from general public abuse		

[*Note*: Throughout items (1) to (10) the contractor will need to ensure that adequate security against damage, pilferage and risk of accident to general public is fully observed.]

(11) *Suspended timber first-floor*

(a) inspection of timbers for wall plates and joists to assess quality (stress grading for latter) and provision for short-term storage on site	visual inspection of materials to meet design specification	
(b) check on consistency of level for wall plates and joist hangers and 'building-in' of latter	possible control check	inspection for placing of wall plates and joist hangers
(c) check on provision for services within joist sections (e.g. pre-drilled) prior to installation		

Contractor	Design Team	Building Control
(d) check joists for spacing and strutting — check jointing at stairwell	visual check for accuracy of joists and stairwell size	inspection to check construction in accordance with Building Regulations
(e) check provision for tying-in to parallel walls		check on lateral restraint provision

(12) Walls up to roof level

(a) continue to check on verticality, horizontal alignment, d.p.c.s at openings and wall ties	visual check on quality of work, inspection of frames for tying-in	check on verticality, wall tie spacing, etc.
(b) continue to check on the installation of lintels, insulation board and cavity retention		check for U values
(c) check quality of timber wall plates, levels and fixing to walls		check on wall plate installation

(13) Roof construction

(a) check on quality of trussed rafters for material and construction		
(b) check installation for fixing, spacings, bracing and tying into gables	visual check for quality of construction	inspection for roof construction and lateral restraint
(c) check adequate laps on roofing felt, batten pitch provision for soil stack		
(d) check tiling for sufficient fixing points and detail at soil stack	visual check on quality of work	inspection of roof finish including flashings at soil stack
(e) check provision internally for loft access		

[*Note*: Through (11) and (12) the contractor will need to ensure that the stability of scaffolding during construction, and in use, accords with safety regulations. Also he must ensure that all materials meet specification requirements and supervise dismantling of scaffolding after sundries.]

Items for supervision (continued)

Contractor	*Design Team*	*Building Control*
(14) *Flooring to first floor and stud walling*		
(a) check grade quality of tongued and grooved chipboard prior to installation		
(b) check fixing of boarding to ensure staggered laps	visual check on quality of work	
(c) check accuracy of setting-out for stud-work		
(d) check quality of timber studding and pre-drilling for service runs		
(e) check rigidity, verticality and alignment of fixed studs	visual check for room sizing	inspection for correct construction
(15) *Carcassing out for services*		
(a) check location of meters, etc. for incoming services		Electricity, Gas and Water Board supply
(b) check wiring runs for power and lighting	check on accuracy	inspection
(c) check pipe runs for gas supply	check on correct provision	inspection
(d) check on hot and cold water services, including all control and drain valves	check for accuracy of provision	inspection for correct installation
(e) check installation of internal sanitation and drainage	check for accuracy of provision as per design drawings	inspection for correct installation
(f) check installation of pipework for central heating system, including header tank	check for accordance with design drawings	
(16) *Installation of service fittings*		
(a) check location and safety of all electrical power sockets, light roses, spurs, switches and consumer panel	check for accordance with design drawings	Electricity Board to connect and check system

Contractor	Design Team	Building Control
(b) check connections to gas-fire/boiler and cooker	visual check on workmanship	Gas Board to connect and check system
(c) check fitting and connecting of all sanitary ware	visual check on workmanship	

(17) *Plastering of block walls, dry lining of studs and plasterboarding of ceilings*

Contractor	Design Team	Building Control
(a) checks on correct use of background coat for plaster and drying time prior to finishing coat	visual check on quality of work	
(b) checks on adequate fixing, correct taping and dry finish to dry linings	visual check on quality of work	
(c) checks on correct application and finishing of ceilings, including cornicing as appropriate	visual check on quality of work	check on items (a), (b), (c), for satisfactory resistance to flame spread

[*Note*: Items (16) and (17) may be reversed either in part or fully to suit service installations, e.g. plasterboard ceilings generally fixed prior to ceiling roses but wall usually plastered after boxes for power sockets and switches are located. Also dry linings often fixed after sanitary appliances are fitted.]

(18) *Installation of radiators to central heating system*

Contractor	Design Team	Building Control
(a) checks on fitting of radiator panels, connection to pipework and fitting of valves	visual check on quality of work	
(b) checks on connection to boiler, including pump	visual check by services engineer and testing of the system	inspection to check accordance with Regulations
(c) check all electrical connections, including thermostats		

(19) *Insulation at ground floor and roof space*

Contractor	Design Team	Building Control
(a) check insulation blanket at roof space to ensure eaves ventilation, check against covering electrical wiring and check installation at cold-water storage tank	visual check for quality of work and provision of roof space ventilation	check for U values

Items for supervision (continued)

Contractor	*Design Team*	*Building Control*
(b) check insulation of water pipes and tanks in roof space		
(c) check insulation to loft access panel		
(d) check thickness of ground-floor insulation board, check electrical wiring runs against risk of overheating. Ensure insulation of central heating and water pipes		check for U values
(20) *Ground-floor finishes*		
(a) checks that screed to kitchen area is reinforced	visual check on quality of work	
(b) checks that battens for hardwood boarding are adequate		
(c) checks on quality of finish to quarry tiling		checks that finishes are in accordance with approved specification
(d) checks on sanded finish to hardwood strip floor		
(21) *Kitchen units*		
(a) checks for quality of finish on delivery		
(b) checks that correct items have been delivered		
(c) checks on accuracy of installation	visual check on quality of work	
(22) *Second finishes*		
(a) checks on all materials and components for correct specification; check on quality of work	visual check on quality of materials, components and workmanship	final inspection of houses. British Telecom lays 'phones to houses
(23) *External works – road and walkways*		
(a) checks on accuracy of kerb alignments to road for drive access and width of carriageway		

Contractor	Design Team	Building Control
(b) checks construction of road, including correct falls to surface-water gullies	visual check on quality of work	check on size of cross-overs from road to drives
(c) check connection to existing road	inspection for smooth cross-over	
(d) checks on sub-grade to pedestrian pavements, accurate location (for level) of electrical and telephone junction boxes		
(e) checks on laying of paving slabs to ensure regular surface	visual check on quality of work	
(f) supervision of turfing of grassed area between walkway and road (this contains stop-cocks, access for drains and emergency access to gas supply)		local authority installs road lighting, road names and traffic signs; paints road markings

(24) *Landscaping and site clearance*

Contractor	Design Team	Building Control
(a) visual check on turfing of front gardens and grading of rear gardens plus erection of fences		checks on accuracy of boundary fences
(b) clears all remaining materials, plant and rubble from site	final inspection for quality of finish	

[*Note*: Sundry items, such as fixing of rain-water goods, external and internal decoration, polishing of hardwood strip flooring, carpeting (as part of this particular contract), hanging of garage doors, final preparation work. e.g. window cleaning, have not been dealt with, since the supervision is almost entirely the contractor's responsibility.]

This example of site supervision for a relatively small contract indicates the diversity of supervisory control. Much of the contractual quality control is carried out concurrently with each process, whereas the external control and inspection is of completed work. It is obviously in the contractor's own interests to provide such an effective internal control mechanism that the external control is only an approval of work done. There is, however, one minor area where some remedial work needs to be carried out under supervision and this is in the maintenance period. This will normally comprise some filling of shrinkage cracks and making good to decorative finishes. For housing, this is normally carried out 6 months after occupation.

Assignment 8.1
Describe in more detail the type of supervision necessary by the contractor in the fitting of sanitary ware for the bathroom of a typical house. Sketch the details of supply and waste connections. [*Note*: You might find it helpful to refer to manufacturers' catalogues in answering this assignment.]

Assignment 8.2
Illustrate the process of installing a sealed-unit double glazed aluminium-framed window, with hardwood sub-frame, into a brick and block cavity wall. Explain the weatherproofing detail between frame and sub-frame and sketch a section through the complete assembly to illustrate precautionary measures against 'cold-bridging'. [*Note*: Window manufacturers' technical information sheets will be helpful in answering this assignment.]

PROJECT
An area of open land is to be developed to provide a two-storey village hall to cater for the following activities:

Internal
amateur drama, staging (demountable), dressing-rooms/showers/W.C.s;
social functions, kitchen/service bar, toilets;
play-groups, nursery areas (upstairs);
youth club.

External
children's playground;
grassed areas and flower-beds, seating;
car-parking (ground level);
pedestrian walkways.

Select an area of land, near where you live, that might be suitable for such a development. Carry out a preliminary site investigation and determine the nature of the subsoil. Select a suitable form of construction, providing a reasoned argument for your choice of structural form and materials. Specify internal and external finishes for the various activity areas. Produce a set of plans and elevations and design a kitchen layout. Detail the sanitary installations to the dressing-room area and the connections at supply and waste positions. Describe the construction of the car-park and pedestrian walkways and specify any associated work. Describe the formation of the grassed areas and flower-beds, assuming some stepped sections of landscape. Where you feel it would help, supplement your project with photographs of actual work of a similar nature.

Index